ANALOGOSAURUS

ANALOGOSAURUS

Avoiding extinction in a world of digital business

By

Simon Waller

simonwaller

First published in 2015 by Simon Waller in Melbourne, Australia.

Cover illustration by Rory Schmitz

Edited by Alexandra Davies

CONTENTS

ABOUT THE AUTHOR

Simon is passionate about creating a better world through digital technology. He believes that technology gives us superhuman abilities when it comes to shaping our future and we need the know how to use these abilities wisely.

Simon may be somewhat unqualified as a technology commentator; the most technical training he has ever undertaken is Advanced Excel. Instead Simon has more than half a lifetime of experience in business. He has held roles in everything sales, business improvement, management and strategy development. He has worked in organisations with 5 employees to 50,000 employees. He holds a Master of Business Leadership, post-graduate qualifications in Futures Thinking and a long time ago he even completed a Bachelor of Commerce.

Simon lives in Melbourne, Australia with his wife Naomi and their two daughters Miah and Poppy. Apart from playing a bit of basketball, Simon's main hobby is his campervan, a 1990 Nissan Homy called Dennis. When he is not camping in the van with his family he can often be found cleaning it, upgrading it and fixing the things on it that he has recently broken.

ACKNOWLEDGEMENTS

I've decided that doing the acknowledgements might in fact be the coolest part of writing a book. When I look back on the 12 month journey that has got me to this point, I am almost overwhelmed with the generosity that has come my way. It gives me great pleasure to be able to thank those people who have supported me over the last 12 months – the book would not have happened without you.

Without a doubt my thanks must firstly go out to the wonderful Kieran Flanagan and Dan Gregory from the Impossible Institute. About eight months ago I bailed them up at a program we were both attending and asked them if they could help with the name of my book. At that point I had written 20,000 words that didn't make a lot of sense and I was struggling with a sense of direction.

That night Kieran and Dan went away and worked on my book idea during their own time and came back the next day with a mock up of a book called Analogosaurus, complete with a pink dinosaur on the cover. It is perhaps a sign of their branding mastery that in that one pencil sketch I immediately knew what the book was meant to be about whilst simultaneously giving me licence to have a little fun with it. Kieran and Dan, every time someone says to me 'cool name for a book', it reinforces your awesomeness.

Kieran and Dan are members of a much bigger tribe that I am extremely proud to be a part of and who I also owe much thanks. The Thought Leaders community, founded by Matt Church and led by Matt, Peter Cook and Christina Guidotti, has been my main support mechanism during the writing of this book. A little over a year ago I made a commitment to write a book within 12 months, though at the time I had no idea what it would be about or who it would be for. The systems and processes that Thought Leaders put in place made it possible for me to achieve my goal. Quite simply this book wouldn't exist without Thought Leaders, so to Matt, Pete and Christina, as well as many others in the community such as my mentor Laurel McLay, Yamini Naidu, Jason Fox and Dermot Crowley, thank you. You guys are my tribe.

Another massive thanks must also go out to my editor Ali Davies. I am not sure whether Ali took me seriously when I came to her without a clear narrative and no solid chapters eight months ago saying that I wanted to publish a book in six months' time. Although we haven't quite achieved my more ambitious timelines it is only due to Ali that we got close at all. It has been a pleasure working with you, thank you for your support and guidance on my writing, for directing me through this process and most importantly for keeping me focused and on track for nearly eight months.

And finally, my deepest and dearest thanks must go out to my family. To my wife Naomi, thank you for your understanding when I took my laptop on holidays, and when I worked on weekends. Most importantly thank you for encouraging me to leave my last real job and set off on this crazy adventure four and a half years ago. What our life might lacks in certainty it makes up for in excitement and purpose. To my two girls Miah and Poppy, this book is ultimately for you. I only hope that by other people reading it, in just a very small way, they will create a better future for the two of you to grow up in. If they do that, it will have all been worthwhile!

A FEW WORDS ABOUT THIS BOOK

This book has been written for people who work in business. This is not to say other people won't benefit from it, in fact I am sure they will. There are whole chapters written about the challenges of technology in our personal lives, and as you will find out later bad technology use in our personal lives is intrinsically linked to bad technology use in our professional lives. So although there is something in here for everyone, this book is about technology in business, and for a few reasons. Firstly, business is what I know; secondly, improving the way we work is the difference I am here to make; and thirdly, despite the relationship between personal and professional technology use, there is a big difference in how we make decisions about technology at home and at work.

I started working in my family's business from about the age of 18. While completing my undergraduate degree I had a part time job helping out my dad and doing the bookkeeping. Since then I have studied business and owned businesses, consulted to them, started them, trained them, worked in them and managed them. This is where I have tested my theories and developed my skills and knowledge.

What my experience has taught me is when used well, digital technology has the potential to improve both the processes and the outcomes of work. I believe that digital technology can amplify our awareness, communication and collaboration, and it can enhance our decision making, which in turn has the potential to create a better society for all of us. This is the difference I can make. I have worked in a business that literally moved a mountain and I want that power to be used well.

There are few boundaries to where digital technology can be found (and for some that is a genuine problem) but not all technology is created equal. The way we approach digital technology in our personal lives is fundamentally different than how we do at work[i]. At work we are much more concerned with actions that provide a competitive advantage – in fact we almost have an obligation to pursue them. In our personal lives we have the option to ignore faster, cheaper more effective ways of doing things for no other reason than because we feel like it[ii].

---

[i] Normally I am not one to make a big distinction between our work time and our leisure time. I much prefer the idea of work-life integration than work-life balance. When digital technology allows you to work anytime and anywhere, why not work from where you want when you feel like it?

[ii] It is perhaps somewhat ironic that many of us have been faster to integrate digital technology into our personal lives than in our work lives, but perhaps one of the reasons that we have been slow to adopt technology at work is because of negative experiences that we have had in our personal use of it.

So, if you are struggling with digital technology at work, this book is for you. I hope it guides you towards a more rewarding digital future.

INTRODUCTION (MY LITTLE RANT)

The first dinosaurs emerged on the earth about 230 million years ago and they could certainly teach us a thing or two about both survival and extinction – survival because they were the dominant land animals for about 135 million years, and extinction because, well, now they're gone.

Around 66 million years ago, there was a mass extinction event in which nearly three-quarters of all plant and animal species on Earth disappeared. Amongst them was almost every species of non-avian dinosaur. Although the exact cause of the extinction is unknown, a 180 km wide crater discovered in the 1990s provides strong evidence to suggest that it was caused by a comet or asteroid hitting the earth.

If this were the case, the dinosaurs would never have seen it coming. They had no satellite-based early warning system to help them detect such threats, and with rather unsophisticated brains (at least compared to what we now possess), they would have had no real understanding of what an asteroid was. The dinosaurs weren't completely stupid, but they were unlucky. Their destiny was outside their hands.

There is another mass extinction coming, this time not an extinction of life but an extinction of jobs. The outcomes might not be quite as catastrophic but they will be just as inevitable and just as widespread: research at Cambridge University suggests that about two-thirds of jobs are at risk of going extinct over the next 20 years.

Unlike the dinosaurs, we can see this extinction coming. The proliferation of digital technology is already affecting our workplaces, it is constantly in the news – in fact, it has even changed the way we access news. We have time to react and do something about it. If we don't survive this extinction, it won't be down to luck; it will be down to the decisions that we make, and a fair dose of stupidity.

Of course, after the asteroid hit 66 million years ago, not everything was wiped out. Based on 75% of species disappearing, simple maths would tell you that about 25% of existing species survived. Amongst these was our distant ancestors. In the post extinction era, many of these species thrived by evolving to fill the void left by the dinosaurs. This will also be the case following the extinction we are about to see, as people with the willingness to evolve and build new skills will survive as well as thrive in the new business landscape.

As they say, nothing lasts forever. The way we work has continually evolved in much the same way that animals and humans have throughout history. As part of this evolution, we have continually discarded old work tools and adopted new ones. Sometimes these were physical tools, but they have also included human tools such as communication techniques or the way we organise work, and increasingly they are digital tools that allow us to share information and communicate. This evolution in the way have risen and fallen based on their comparative use of tools and technology.

At an individual level it may not be so obvious, but the same forces are at play, especially in a work environment. The ability to use our tools well (whether they be physical, human or digital) can improve our performance and provide a competitive edge when it comes to promotion or future employment.

And this now brings us to the digital age. We are seeing a new generation of tools that provide a level of advantage like never before. They are providing benefits that 50 or even 20 years ago we couldn't have imagined. And unlike the changes in technology we have seen in previous eras, this one is happening really fast. We are no longer measuring change in millennia, centuries or decades – we are now measuring it in years or even months.

These digital tools are so powerful that we are redesigning our other work around them. The digital age has allowed human and physical work to be increasingly represented in lines of code. The design of a car is now stored in computer files rather than on paper or in the head of a master craftsman. Communication and collaboration are now enabled as much by technology as by human endeavour. And although we may complain about digital technologies encroaching on our lives, the continued growth of these technologies reflects the distinct competitive advantages they offer.

Digital tools are quickly outcompeting their analogue alternatives. And like the generation of tools before them, analogue tools are quickly going the way of the dinosaurs. They are quickly becoming extinct.

Whether it is a result of ignorance, fear or laziness, many of us will find it difficult to let go of our analogue past and embrace our digital future. We may make excuses to ourselves so that we can avoid the inevitable change. We might say things such as, "I can remember more when I write things down", or you might claim that you like the 'feel' of pen on paper. But the reality is that we live in a world that is being digitised, being captured as information. Information has become the lifeblood of business and in this world our analogue tools can no longer compete.

This digital shift is no longer restricted to a particular industry or job role. Although it will undoubtedly impact some more than others, there are very few people, if any, that won't be affected by this change. Our individual choices don't really matter, because the digital revolution is as inevitable as the Industrial Revolution, and every other revolution that has come before it. The potential for improvement is so great, the opportunity so big, that no one has the power to stop it.

So here's the hard sell: your future is being shaped by your customers, your suppliers, your colleagues and your competitors. Some of them are already developing and implementing new tools, and they will set the new standards that everyone else will have to meet, or exceed, if they want to maintain relevance in a world of digital business. When it comes to the crunch, you just won't be competitive if you don't.

You may not be able to shape the future, but you can still influence your destiny. You get to choose if, when and how you join the digital revolution and this book has been written to help you make that change.

As they say, the Stone Age didn't end because they ran out of stone, and the analogue age won't end because we don't have enough paper. The question isn't whether your analogue tools will go extinct. The only question is: will you go down with them?

SO WHAT'S THIS ALL ABOUT THEN?

A WORLD OF DIGITAL BUSINESS

At the turn of the millennium I found myself sipping champagne in the south of Spain. I was living in Nottingham, England at the time but along with a small group of friends had decided to escape the UK and spend this once in a thousand year event in a small café overlooking the marina in Porta Banus.

I remember at the time there were two stories that dominated the news headlines for months leading up to this point. The first was the lavish New Year's Eve celebrations that every bar/restaurant/nightclub in England seemed to be planning. The second was the looming catastrophe of Y2K, a computer bug that had the potential to create worldwide chaos by decimating computer systems around the world. We had gone to the south of Spain because of our desire to escape the first one, but it probably wouldn't have been a bad place to be stuck if the second one turned out to be true.

Ultimately, like the celebrations of the many people who paid extortionate prices to celebrate the millennium in UK nightclubs, the Y2K bug turned out to be a bit of a fizzle. As a result it has largely been relegated to a small, dark, rarely accessed corner of our memory. In case you have forgotten (or are too young to remember), Y2K or the Year 2000 Problem was a concern that all of our computer systems were going to fail at the dawn of the new millennium. Although technically the new millennium wasn't due to start until 1 January 2001, neither the people of the world, nor the computer programmers who shared in over US$300 billion of preparation and remedial work as a result of Y2K, seemed willing to wait another year.

The problem seemed to be that many of our early computer programs were written using two digits rather than four to represent the year (1/5/99 instead of 1/5/1999). The reason for this was that when these programs were written, we were so short of computing power that we couldn't spare enough hard drive space for an extra couple of digits. As a result there were significant concerns (significant enough to spend US$300 billion) that on 1 January 2000 a whole bunch of computers would think that it was 1900 and either stop working or create more problems for us than usual.

As it turned out, public transport didn't break down and ATMs didn't start spitting out money. Even the small business sector, where a huge proportion of businesses either couldn't afford, didn't understand how or couldn't be bothered to deal with Y2K came out the other side unscathed. On the first work day of the new millennium the US Commerce Department's Y2K Help Center for Small Business received about 40 calls, pretty much the same as on any other day.

There is much debate as to whether the money spent on Y2K was well-spent. There was minimal disruption, even in the businesses and countries that had spent little or no money preparing. As a result it has since been called the 'hoax of the century', but clearly most people, whether in business or government, weren't really sure.

In some ways Y2K is a perfect example of everything that was wrong with digital in 2000. It instilled us with fear, it was hugely expensive, very few people understood it and it seemed hard to get independent and trustworthy advice from those that did. Digital was fairly well-entrenched in organisations, mainly as a result of Microsoft Office and email, but quite frankly the whole thing was fairly underwhelming.

A few months after surviving Y2K I headed back to my home town of Perth, Western Australia. My first job on return to Perth was working with my dad as a commercial boat broker at Salter Marine, which was like a real estate agent for commercial fishing boats. We had an office on the fishing boat harbour in Fremantle, and although we had one of the first websites in the industry (which Dad had created and maintained himself), this was still very much an analogue business.

I did have access to digital tools; I had a laptop on my desk and I used it to send the odd email, write up contracts and create spreadsheets, but all this digital work was done behind the scenes. Networking was done face to face on the wharf, deals were made on handshakes, contracts were sent via fax machine (or printed and posted for those who preferred the old ways) and people paid by cheque.

Almost 10 years to the day after I started working at Salter Marine, I left what my friends would describe as my last real job[iii]. I had been working as part of Rio Tinto Iron Ore's Scenario Planning and Strategy Team in Perth but my wife Naomi and I had decided to move our family to her home town of Melbourne on the other side of Australia.

When I left Perth in 2010, business was distinctly different than when I had arrived. At some point in the first decade of the 21st century there was a fundamental shift away from analogue towards digital technology. This shift had been a long time coming. Technology purists could probably trace its roots back to a time

---

[iii] By 'real job' I mean a job where you trade a set number of hours per week for a set amount of money, in other words, a salary, but I do not mean that work in salaried positions is more 'real' than other types of work.

before desktop computers, but for most of us it started somewhere around 1982, *Time Magazine*'s Year of the Computer. At that point change was more of a trickle, but after 2000 this trickle suddenly became a flood.

According to Dr Martin Hilbert and his team at the University of Southern California, 2002 was the year that we moved to having more of the world's information stored in digital formats than in analogue ones[3]. In 2003 the Blackberry smartphone was released, freeing email for the first time from the confines of our office. Then in 2007 we got the first iPhone, a device originally designed for the consumer market that subsequently laid waste to the other big phone manufacturers of the time, including Nokia and Motorola. And if that wasn't enough, in 2010 we got our first mass-produced tablet computers in the form of the iPad.

By the time I left Rio Tinto and moved to Melbourne, business had become digital. Networking was starting to happen on LinkedIn rather than face to face, meetings were increasingly conducted using teleconference software inside organisations, or on Skype outside of them, and the expectation was that documents were emailed rather than posted.

This may not seem dramatically different to 10 years earlier. The processes were still largely the same, but the medium they were conducted in was different. Over that decade, starting in larger organisations but quickly filtering down to smaller ones, it was increasingly expected that the output of work and the medium for exchange was digital. When I worked at Salter Marine there may have been digital parts but it was largely an analogue process; by 2010 the processes were now largely digital with maybe a few analogue bits left.

Somewhere in the first decade of the 21st century we entered the world of digital business. And we're not going back.

THINGS ARE SPEEDING UP

For many, the challenges of digital technology are more visible than the opportunities. This is because our established and familiar work systems are being overwhelmed with the increasing flow of information that digital technology brings. We persist with out-dated processes for managing information because we don't know a viable alternative and it seems easier to just keep doing what we have always done. Once our systems are overwhelmed, once our information tanks have reached capacity (and regardless of the quality of value of the information in them), any new information spills over the top and is lost. The impact of this often extends beyond the workplace and its normal working hours. Our inability to keep on top of it means that it flows into our downtime and can negatively affect our work-life balance and our wellbeing.

Information overload is commonly seen as one of the biggest challenges facing people in business today, and it is a direct result of the increasing volume and increasing complexity of information flows. It is estimated that in the 21 years between 1986 and 2007, the amount of information we received every day grew from the equivalent of 55 newspapers to 175 newspapers[4]. In that time we also saw a growing spread of information channels so that as well as newspapers, mail, phone, TV and radio, we have had to incorporate a host of new communication methods such as the internet, email, social media and SMS.

One of the results of this information overload is increased time pressure. More time is required to juggle the different information channels and to process the quantity of information. The pressure is on to respond quickly to others as communication becomes increasingly real time. As a result, many of us are working long hours, not in order to manage our workload, but to manage our information load. For some, working late or on weekends when there are no meetings scheduled is seen as the only way to get on top of the information flow that can easily sweep any of us up in its tide.

Apart from managing the impact on work-life balance, it has also become incredibly difficult to maintain focus on core tasks. This is in part because of information overload, and in part due to the distracting nature of many digital tools. The challenge of balance and focus has grown rapidly over the last decade. Since the introduction of mobile email services by Blackberry in 2003 and later the iPhone and other mobile technology, information has escaped the confines of our office and seeped out into our personal lives. Combine this with the distracting and attention-sapping nature of many digital tools (often considered a design feature by developers) and we have the recipe for an attention deficit disaster.

This competition for your attention is described by American economist Herbert Simon:

> *...in an information-rich world, the wealth of information means a dearth of something else: a scarcity of whatever it is that information consumes. What information consumes is rather obvious: it consumes the attention of its recipients. Hence a wealth of information creates a poverty of attention and a need to allocate that attention efficiently among the overabundance of information sources that might consume it.*[5]

It is little wonder that digital technology can be overwhelming for many users. And it is probably unsurprising that the reaction is often to push back, limit our technology interactions and try and reduce the impact it has on our lives. But regardless of our individual choices, most of us have been unable to stem the tide: technology use and information flows continue to grow regardless. So many of us are not only in a relationship with technology that is overwhelming, but also feel like we are in a position of helplessness.

It is the type of abusive relationship that our parents would probably warn us away from. Our partner doesn't get us and we certainly don't seem to get them (often it seems like they are just speaking a different language). We think we are pushing the right buttons but often they just shut down, or freeze up when we're in the middle of something important.

We used to be able to escape and find a quiet place away from home or the office where technology wouldn't be able to find us. We thought we had set clear boundaries, time for us both to be alone, but it wasn't respected. Now it just seems like we're being stalked, not in a nice, friendly ex-girlfriend Facebook-ish type of way but in a sinister, I'm-keep-tracking-of-your-every-move-and-may-demand-your-attention-at-any-time type of way.

Technology made some big promises to us when we got together. It was going to do wonderful things for us, make our lives easier. But it now seems that some of those promises (especially the one about hoverboards) were a bit hollow. We have worked hard at this relationship, we have invested hours in trying to make things work. Every time technology beeps or vibrates, we drop everything to attend to its needs. You would think that with this type of commitment, and after 20 or more years together, our relationship with technology would be better, but for many of us it is just getting worse.

And it is about to get much worse. All these challenges pale in comparison to what is really at stake: our future relevance.

Technological unemployment refers to employment caused by changes in technology. This type of unemployment is not new but it has certainly become more prevalent since the Industrial Revolution. The crux of technological unemployment is that as we create new tools and technologies to make tasks easier, we displace the workers who used to do that work by hand.

For example, agriculture accounted for nearly 90% of all US employment in 1790; by 2010 this had fallen to about 2.2%. Over the preceding 200 years, technological innovations had slowly displaced farm workers, yet since World War II alone farm production has increased by 250%.

So what happened to all the workers? Traditional economic theory would suggest that they reskilled and got new employment elsewhere – that technological advancement reduces the costs which lowers the prices and this means that society has spare cash to spend on other things. Then, the production of these other things creates new employment opportunities that didn't exist previously.

When the Luddites protested against the introduction of spinning machines in 1760 it was estimated that in England there were 5,200 spinners using spinning wheels, and 2,700 weavers – in all, 7,900 persons engaged in the production of cotton textiles. At the time, spinning machines were opposed on the grounds that they threatened the livelihood of these 7,900 workers. Yet in 1787 – a full 27 years after the invention appeared – a parliamentary inquiry showed that the number of persons engaged in the spinning and weaving of cotton had risen from 7,900 to 320,000, an increase of 4,400%[6].

Since the Industrial Revolution, the general view of technological unemployment has been based on the above thinking – that although technology may create a jobs churn, people will be re-employed elsewhere in the economy. But more recently there have been murmurings that this may no longer be the case. Reasons for this are detailed in books such as *The Second Machine Age* by Andrew McAfee and Erik Brynjolfsson and *The Lights in the Tunnel: Automation, Accelerating Technology and the Economy of the Future* by Martin Ford. Essentially, technology is now changing at such a rate that displaced workers cannot be retrained and redeployed fast enough. It is as if the churn is happening so fast that the cream is flying out of the bowl before it can be turned into butter. As the rate of technology change is still speeding up, it would seem likely that this is a growing rather than a shrinking problem.

Although both of these books are well worth reading, they both take a rather macro and somewhat philosophical view of this change. At a macro level it is easy to be seduced by the broader social and economic benefits that technology provides, but at an individual level it is still everyday people losing their jobs.

Exactly how many people? Research conducted by Carl Frey and Michael Osborne at Oxford University[7] suggests that up to 47% of US jobs are at high risk of automation over the next two decades and an additional 19% are at medium risk of automation. This means that there is a fairly good chance that two out of every three jobs will be automated during the next 20 years.

Take a moment to let that sink in: 66%.

Some of these people may never find employment again. Some may take months or years to find employment. Some lucky others will have a transferrable skill set and find new employment fairly easily.

Ultimately it doesn't matter if you are part of the 66% majority or the minority. Every single job is going to be impacted by digital technology in some way. As discussed earlier, digital is not going away, it is an unavoidable part of our future. Some of us have managed to sidestep and shimmy around in denial for the last decade but eventually it is going to catch up with us.

At the moment many of us are making excuses to avoid acting. We pretend that we have some special skill set that technology is just not capable of replacing. Or that we have unique characteristics critical to our particular role that cannot be replicated. This may or may not be true, but that may not really matter. It is not about digital being better in every single way, it is just about it being better enough.

There is a legitimate argument that reading books on an e-reader or a tablet is not as enjoyable as reading on paper, but this hasn't stopped the growth of digital books. Way back in 2011 the sales of digital books on Amazon surpassed that of printed books. This is not because digital books are better in every way than paper books, but they are better enough. They are cheaper to produce and cheaper to distribute; there is no such things as over-runs and unsold stock; and they are highly portable. At the moment I have about 50 books on my tablet device. Each one of these I can search, highlight, and copy text from. Sure, sometimes the layout isn't quite right, I can't hold it as easily in one hand and the backlight display is not as easy or enjoyable to read, but the benefits outweigh the costs and it is better enough for me to use it in place of printed books.

The fact that eBooks are better enough is fairly easy to see as a consumer and it is reflected in the extraordinary sales growth of e-readers and eBooks over the last few years. But sometimes it is more difficult to see as the producer. To get a different perspective let us for a moment take this discussion out of the present and go back to another revolution in writing technology, the shift away from carved stone tablets to papyrus, around 2500 BCE.

It would have been incredibly easy for the stone masons to dismiss the significance of papyrus. In so many ways papyrus was an inferior product to stone. It was extremely fragile, couldn't be stored in conditions that were either too wet or too dry, wasn't naturally occurring and required an extensive manufacturing process. You could almost imagine the stone masons sitting around at the end of the day discussing this newfangled technology and feeling pretty safe that they were working with one of the most durable, abundant and cheap resources in the world: rock.

But papyrus didn't need to be better in every single way to displace, and ultimately replace, stone carving. Compared to rock it was incredibly portable, and even though the manufacturing process was more complex it would have been fairly fast to create a new document on papyrus compared to on stone. Yet the stone masons might have initially felt a certain level of security because they were working with a far more enduring product. Maybe they would have got into the papyrus business if the retooling and retraining costs weren't so high and they hadn't invested so heavily in their current skills and tools...and also if they weren't slaves.

But ultimately they would have seen their relevance starting to wane. They were probably still getting good work doing monuments but some of the public communication work started moving onto papyrus. Maybe later still Pharaohs started to get buried with scrolls rather than just having all their conquests carved into stone. Fast forward a few thousand years and stone masons have become the domain of novelty and nostalgia. Most of the work these days is in tombstones, and everyone knows that's a dying business [insert groan here].

In every technology revolution there are winners and losers. In the Industrial Revolution the biggest winners were the new industrial class that built wealth and power on the back of machines and cheap labour. But they weren't the only winners – the next best off were those that actively engaged with change. They saw opportunity in industrialisation; they realised that the running and administration of industrial organisations and industrial equipment would require new skills and abilities, and they sought to fill this vacuum.

The losers were the artisans, the guilds and the Luddites who focused on what they had to lose rather than the new opportunities that were available. The digital revolution will be no different. We have already seen unbelievable riches accrue to the new emerging technocracy. Substantial benefit will flow to those that proactively engage with the opportunities that digital technology offers and, as always, the losers will be those who focus on what they have to lose, in this case the digital laggards.

At a macro level we can point the finger of responsibility at government, unions, academics and others to do something about the inevitable shift towards technology. But on a practical level, the problem is ours. Our future relevance is at stake and it is up to us to do something about it.

WE HAVE CHANGED SLOWER THAN TECHNOLOGY

I assume that if you are reading this book, you already harbour some concerns about the impact of technology on your work. How did we get into this predicament? How did we arrive at a point where in the next 20 years, two-thirds of our jobs will be automated, and yet most of us have done little or nothing to prepare ourselves for it?

Over the last couple of decades, both the variety of digital tools that we use and the amount of time that we spend using them has continued to increase. Computing power has been increasing at a fairly steady rate for more than 40 years. In a 1965 article for *Electronics Magazine,* instructively titled, "Cramming more components onto integrated circuits", Gordon Moore, co-founder of Intel Corporation, described a trend that suggested the number of transistors that could fit on a computer chip would double every two years. In more practical and understandable terms, Moore's Law is often stated as 'computer power doubles, or halves in cost every 18 months' (even though this is not exactly what Gordon Moore wrote).

Although this sounds somewhat simple, the repercussions of it are staggering. As pointed out by Ray Kurzweil, with linear development we would achieve 100 years of progress in the 21st century, but with exponential growth we will achieve more like 20,000 years of progress (at today's rate) over the same timeframe[8]. This is the law of accelerating returns.

So if we think back to a time in the early 1990s when many of us started our journey with digital technology, the water in the pot was only lukewarm. Most of us didn't understand the future impact it would have and our approach was commensurate with this. In those early days, digital technology was expensive, slow, unreliable and hard to use and for many of us these early experiences with digital shaped our habits for years to come.

A 1992 Windows laptop would have set you back $3,000 or more and had less computing power than today's basic smart watch. If you were one of the fortunate few who had access to the internet, and assuming that no one wanted to make a phone call at the time, you would set your modem to connect and head off to make a cup of tea. If you were lucky, you would be connected by the time you got back.

This was a time when we were plagued with computer issues. We dealt with the 'blue screen of death' and documents needing to be saved manually. One inadvertent button push could destroy documents, programs or even your whole computer and when things did go wrong we were confronted with obscure error messages written in another language.

This was the environment that many of us forged our relationship with technology in. So like a rat trained to avoid electric shocks, we were conditioned to avoid technology. And even though digital technology was improving at an astounding rate, when it was only half or even a quarter as frustrating as before, it was still frustrating.

It is not surprising that many of us developed a take it or leave it (but generally prefer to leave it) approach to technology. We had little islands of digital in a sea of analogue but mostly this was a hindrance rather than a help. In general it was easier to convert our digital into analogue than work in two different systems. We would print our emails to read them, and although we might have had an electronic calendar we would carry a separate paper diary to record things in. There is a good chance that we would also print out a copy of our electronic calendar each day just to be sure that we had a reliable copy with us. If we were senior enough, we might have had an executive assistant to print out our calendar for us and carry out our other digital tasks, and if we weren't senior enough then we were probably dreaming of the day that we were (I write this in past tense but am acutely aware that this is still the present for some.)

It is for this reason that the original predictions of the paperless office turned out to be misplaced. A 1975 *Businessweek* article titled "The Office of the Future"[9] predicted the growth of digital tools would dramatically reduce offices' reliance on paper, largely rendering it redundant by 1990. But instead quite the opposite happened – during the 1980s and 1990s paper consumption boomed. Primarily this was because with digital tools we could create new information at an unprecedented rate, but when it came to sharing, transporting, storing or otherwise using the information, it still needed to be printed so that it fit into our analogue systems (interestingly, as we moved from a world of analogue business to a world of digital business in the early part of the 21st century, office paper consumption in developed economies plateaued and it is now in decline).

So although digital was increasing in potential at an astounding rate, most of this went unnoticed. It only impacted the very small part of our jobs that was digital in nature, and even then, only for as long as it took to turn whatever we were doing from digital into analogue.

In this baptism of fire most of us put together a fairly simple and stable digital arsenal of tools. Of course, we didn't get to choose them – that was someone else's job. So at the same time as we were trying to avoid technology, we were also being conditioned to be passive rather than active consumers of it.

In a business environment dominated by specialisation, we are required to defer to those who know best rather than ask questions. And in the case of technology, this has been the IT department. Thankfully, *not* asking questions of the IT department has often been an easy out. Although human beings like the rest of us, they seem to speak a different language, and so instead it has been easier to accept whatever device is dropped on our desk and use it, or not use it, to the best of our abilities rather than call the help desk and listen to someone speak in an alien tongue.

And if you think the IT department is frustrating, you are probably much worse. Our help desks are manned by highly skilled and intelligent computer scientists who in their own time build smartphone apps and hack NASA's computers, but in daylight hours they are reduced to telling people to restart their computers. To reduce the 'pain' on both sides of the equation, IT departments everywhere have tried their best to keep things simple. They have provided you with largely the same laptop (Dell, IBM, HP anyone?) with largely the same operating system (Windows) with largely the same software (Microsoft Office) for as long as you can remember.

What we've been getting is the Model T Ford of digital technology. You could have the Model T Ford in any colour you liked as long as it was black. And why was the Model T Ford only available in black? Because black paint dried faster. It wasn't for the consumer; there was no product testing that identified black as the preferred colour. It was purely because it suited the people on the production line. And this is exactly what has happened with your digital tools. They aren't designed to suit you, they are designed to suit the IT department that has to manage them.

As a result we have trained our users to believe that the only useful technology is a PC laptop with a 15 inch screen. Not because it is the best, but because it is the only thing they have ever been able to try. Users are generally not encouraged to provide feedback on their experience. The hardware is bought on price, the software is predetermined and the only real option provided to the user (if they're lucky) is whether they want a backpack or briefcase style laptop bag. We might have thousands of different job descriptions in our organisations but we still have one standard operating system.

When IT departments were first formed to support digital accounting packages in the 1980s, the mandate of stability, security and price was appropriate. And back then, if you were provided a desktop computer (or 10 years later a laptop, or 10 years later than that a Blackberry) you didn't complain, you just counted yourself lucky that you got one at all – not necessarily because you wanted to use it but because of what it symbolised in terms of importance and status.

Some of us have started tinkering around the edges of this digital explosion but mostly we do it in our personal time, and largely without thought for its business application. We keep in touch with friends on Facebook and other social media platforms, we download apps, buy new smartphones and tablets and may have even invested in a smart watch or fitness tracker. Somewhere around 2008 the best technology was no longer being used in business, it was at home. But technology within our organisations has remained the domain of the IT department. The checks, balances and outdated KPIs of IT departments has meant that most of them can no longer keep up.

Most of us seem unaware, unwilling or unable to do something about this. We may indulge in digital technology at home but go back to doing the same old things at work. We might acknowledge that Facebook is great for keeping in touch with family and friends scattered across the globe and then go to work and use email to communicate with our suppliers and customers. We might expect our personal photos taken on our smartphones to automatically appear on our home computers but be willing to accept that we can only access our work files when we are sitting at our desk and connected to our shared drive.

Given all this, perhaps we can understand why many of us seem to be trapped in a time warp, using 1990s technology over and over again. 1990 might not seem like a long time ago (and in the context of all human history it really isn't) but in the context of digital history it is significant. Between the 1950s and the 1980s, computing power increased 100,000 times. Between the 1980s and 2010 it increased 100,000 times on 1980 levels, meaning a 10 billion fold increase over the 1950s. It is now increasing more each hour than it did between 1900 and 1990[10].

This explosion in computing power has led to an explosion in digital choice. Improvements to automobile production have led to shorter development cycles, increased options and improved performance, and we have seen the same thing with digital technology. Even though they were produced 14 years apart you would be hard pressed to tell the difference between a 1909 and a 1923 Model T Ford (even though there were three different variants during this time), just as you would be hard pressed to tell the difference between the first few generations of Microsoft Windows. Windows 3.0 was released in 1990 and the interface didn't significantly change for five years until the release of Windows 95. In that five years, the product was updated twice. In comparison, Microsoft's cloud-based Office 365 service is updated every three months.

There are perhaps few better examples of the extraordinary growth of digital technology than that of driverless cars. In 2004 DARPA, or the Defense Advanced Research Projects Agency, offered a US$1 million prize for whichever organisation could create an autonomous vehicle that could cover a 150 mile (240 kilometre) route in the open desert. In the first year of the challenge not one vehicle finished, and the best performing vehicle by Carnegie Mellon University covered just over 7 miles.

The following year the prize money was doubled and five vehicles managed to finish the course, the fastest one at an average speed of under three miles per hour. In 2007 the course was moved to a more challenging urban environment format (but thankfully hosted on an air force base and not in suburban streets). That year the winner covered the course at an average speed of 14 miles per hour.

A few years later the same team, now working for Google, started road testing driverless cars in Nevada and later California. In 2012 Google posted on You Tube a video of Steve Mahan, who has lost 95% of his vision and is legally blind, being transported from home, through a drive-through restaurant, to the dry cleaners and back home again, in a car without a driver.

In 2014, Audi raced an autonomous vehicle against an identical vehicle with a human driver around the famous Hockenheim track in northern Germany. Reaching speeds of up to 190 miles per hour (305 km/h), the driverless car won the race by five seconds.

The approach we developed to digital technology 20 years ago is now out of date. At the time, avoiding technology use was quite a good strategy for improving your productivity and general performance. The cost and time invested in hardware and learning would have taken a long time to recover, the frustrations of technology would have distracted you from more important things and there was a lot of time lost converting analogue information to digital and back again. There is a good chance that back then we made some grand statements: "I will never use an electronic calendar", "iPads are just media consumption devices", or maybe, "pen and paper help me think".

But in the present day, our avoidance techniques lead us to leave tablets sitting uncharged on our desks, serving only as extraordinarily over-engineered paperweights, or to pass them to the kids to play games (or do whatever kids do with technology). We may feel trapped by our previous declarations, because we committed to our paper systems when they were the best things going and we are determined to stick to them even though now they are not. Unfortunately, we have held on to this out-dated approach for too long.

The truth is the equation is now different. Digital can no longer be avoided even if you want to, and frankly, you shouldn't want to. Digital is making the way we work cheaper, easier and more flexible. We can have a level of reliability and information backup that we are unable to reproduce in our paper systems. And there is so much upside yet to come.

We will still have moments of disenchantment where technology doesn't do what it's meant to (or at least, we can't get it to do what we want), but these events are becoming rarer and rarer. I am sure we can all think of a recent time where technology failed us – an app closed unexpectedly, we lost someone's contact details, we couldn't find an email – but if you think about this in the context of how much our lives are now intertwined with technology, these events are now fewer and further between.

It took 20 years for our organisational culture of paper use to catch up with the technology of the PC, and starting with email in the mid-1990s it is now 20 years since the mass availability of digital technology. It is time to stop pushing back. Digital technology is accelerating away from us and if we don't get on our bikes and start pedalling, we won't ever catch up.

IT'S OUR PROBLEM AND WE NEED TO FIX IT

Take a moment to look around you. Soon these will be the good old days, the days before digital technology *completely* changed the workplace. At the moment, digital is just seeping in through the cracks, but the cracks are widening and the flow is getting stronger. Eventually the flow will build into a cascade of digital technology, carving new channels, and reshaping the landscape of our working and personal lives. Although making predictions of the future is fraught with danger, one thing we can be fairly confident about is that in our lifetime we will never have less digital technology than we have right now.

This is as low tech as business is ever going to get.

We might have moments where we wish our lives could be technology-free, but this is never going to happen. We may even consider enrolling in a 'digital detox' or some other program that advocates a life without technology, but in reality these programs don't even fit the criteria of a Band-Aid solution. If we are going to live in a world of ubiquitous technology, avoiding it is not an option. 'Learning to live with it' is not an option either. The only real option is to embrace it. We need to understand its strengths and weaknesses, learn how it works and develop the confidence to apply it when we need to. Ultimately we need to develop the judgement to know when to use technology and when to leave it alone.

For a long time we have put the development of our digital knowledge and skills in the hands of others. We assumed that someone would give us the best technology to do our job and if it was important they would show us how to use it. It is a reasonable conclusion to make – surely if the benefits of you being more effective accrue to your organisation, they would want to help you be more effective in any way they can. But although the benefits might flow to the organisation as a whole, they generally don't flow to the people who make decisions about what technology to give you, and they generally don't flow to the people in charge of training and development. Most organisations' learning and development is run just like IT: operated as cost centres with KPIs based on reducing the cost of the services they provide. As a result these teams don't get rewarded for providing better education and learning opportunities, or more productive technology options, only cheaper ones.

In addition, there is a very good chance that the experts who provide you with technology and training don't really know what you need, or even necessarily what all the options are. It would be almost impossible to keep abreast of the plethora of digital platforms and tools that are now coming to market. So instead, the experts focus on some generic tools that do a little bit of what most people need and train you on using that (or more likely, assume you already know how). As a result, many of us are now walking around with the same old technology, using it badly and wondering what all the fuss is about.

But before we grit our teeth and start shaking our fists in anger, we need to reconsider our conclusion about who is responsible for our relationship with technology and ultimately who has the most to gain. The real benefit of digital technology accrues to the individual – we will ultimately benefit the most from the digital knowledge and skills we develop by becoming more employable and better at our jobs. We are the best judges of our own limitations and if we need training to address them, then *we* need to ask for it. For a long time we haven't seen our lack of digital expertise as a limitation in this way; we looked across at the person next to us and saw that they were doing fundamentally the same thing as us and thought it was okay. Maybe it was okay when we didn't live in a world of digital business, but now we do.

In psychology there is something known as the Dunning-Kruger effect. This states that people who know a little overestimate their ability and think they know a lot, while people who know a lot often underestimate their knowledge relative to others. For a long time we have known a little and thought it was a lot; eventually we need to realise that even if we already know a lot more than we used to, there is still so much more to learn.

At the end of the day, this is our problem and we need to spearhead the campaign to fix it.

So what would it take for you to embrace technology? Not just bear it, not just like it, but truly embrace it. Because unless we can do this we face an uncertain future. A future of increasing irrelevance as digital technology replaces increasingly complex cognitive tasks. A future of increasing anxiety as work speeds up and information flows continue to grow. A future of increasing vulnerability as digital technology enables the separation of work into bite-size chunks and dishes it out to part time and contract workers at the expense of full time employees.

For you to change, there are three things that you might need. The first thing is to know that this matters to you, this is about *your* job and *your* future, and not someone else's. The second thing you might need to know is what's in it for you. What would you get out of fully embracing the opportunities of digital technology and investing time and energy into learning how to use it? And thirdly, I think you need to know that it is possible, and ideally that it is easy – that there is a process you can follow to get from where you are now to where you need to go with the least amount of effort.

And this, my friends, is the focus of the rest of this book. The next section is titled 'Relevance' because it is all about why digital disruption matters to you. It discusses the changing nature of work and the impact of using digital technology on the way we operate. It looks at the process of value creation in our jobs and how we can use our digital tools to improve it. The following section, 'Reward', is all about value. It looks at the different value opportunities that digital technology offers and how you can tap into them. The last section is titled 'Realistic' and is all about change. This section looks at the challenges that we have had engaging with technology and the need for a new approach. It answers the big questions such as why kids are so good with technology and what we can do to learn better.

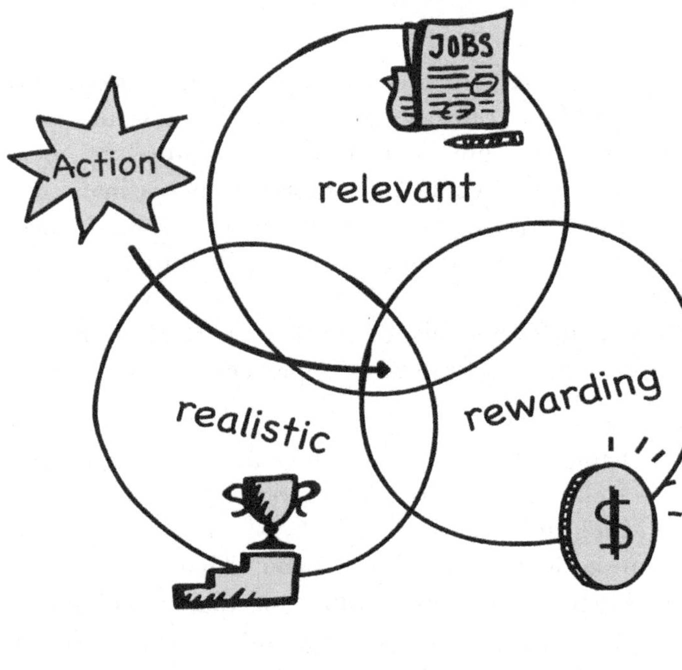

RELEVANT

WHAT DOES IT MEAN TO BE A PROFESSIONAL?

Not long ago I was travelling interstate for work and caught a taxi from my hotel to my client's office. To my surprise when I gave the taxi driver the address he reached under the passenger seat, pulled out a well-thumbed copy of the street directory and proceeded to look up the address.

As I sat in the back of the taxi I tried to remember the last time I had seen a taxi driver use a paper-based street directory instead of a GPS. Only a few years ago GPS units were expensive to buy and difficult to update, though many taxi drivers used them anyway because they were a 'tool of the trade' (and incredibly useful). In a sense it was no different from a carpenter who bought a powered drop saw or rechargeable drill to replace their hand tools.

It is not that a GPS is required to be a professional. It could be that he was such a professional he knew the city like the back of his hand, that he didn't need the map at all as he could quickly and easily calculate the best route to my destination in his head. But this wasn't the case. My driver was using technology but it was outdated, which meant he didn't know enough not to need any technology but wasn't willing to invest in the right tools. I immediately started to question his professionalism. How was he going to know the best route to the destination and was I going to pay a higher fare because he didn't know where he was going?

It got me thinking. Which of these represents the pinnacle of professionalism – having the knowledge, or using the right tools?

In London, Black Cab drivers need to pass a test called 'The Knowledge' to get their licence. The Knowledge tests them on 320 common routes, 20,000 landmarks and 25,000 streets in central London. It is an extraordinarily challenging testing process and it has a 65% drop out rate. The roots of this licensing regime can be traced back more than a hundred years to the Great Exhibition of 1851. Legend has it that the thousands of visitors to the city were so frustrated with the ineptitude of the cabmen that they demanded a better licensing regime. By 1884 there were records of close to 2,000 cabmen applying to take the "examination as to the 'knowledge' [of]...principal streets and squares and public buildings"[11].

Without a doubt, both the commitment to learning and the detailed knowledge of London city streets is a reflection of Black Cab drivers' professionalism. In a pre-GPS world this was incredibly valuable information to hold, but in a digital world this knowledge is rapidly being devalued. A sat nav unit can 'learn' all the landmarks and streets in London instantaneously (compared with an average of 34 months for a Black Cab driver) and at effectively zero cost. And although it is arguable that Black Cab drivers might currently be able to calculate a better route through the city, this is also under threat. Modern sat nav units are using other sources of information to enhance the process of navigation.

A number of mapping programs now provide access to live traffic data that show where traffic is moving quickly and slowly in real time. This data is calculated by collecting data from GPS-enabled smartphones along a specific stretch of road. By identifying smartphones that are moving in a pattern similar to car movement, it can calculate average traffic speed and hence congestion.

So as well as being able to see the traffic conditions in your direct vicinity, you can access the traffic conditions in the vicinity of every other user at the same time. In 2013, when Google purchased the crowdsourced traffic and navigation app Waze, it had close to 50 million users around the world[12]. If you are using either Google Maps or Waze you are tapping into the traffic information of more than 50 million other drivers sharing their traffic information with you in real time. This data is automatically applied when calculating which route to take and can help avoid traffic congestion and accidents.

So the question is, can one driver do a better job of understanding traffic conditions and calculating the best route than 50 million other people and a decent algorithm? Of course the true professional wouldn't choose between these two options, they would want both the knowledge to understand what they are seeing around them and the tools to amplify their performance.

These days you could be an okay taxi driver if you had an ounce of customer service and access to a decent sat nav. Perhaps you could be a great taxi driver if you didn't have a sat nav but had the right knowledge. But in the 21st century, to be a truly professional taxi driver you need to use both.

This is no different than what we would expect from a plumber or an electrician. We would expect them to have an expert knowledge of their subject matter as well as be competent with the tools of their trade. But when it comes to people who do thinking work, we disproportionately value knowledge over the use of tools. There is a general view that if we have the smarts to complete an undergraduate (or postgraduate) degree, or if we've obtained 30 years of industry expertise, then we are professionals. Is it perhaps misguided to consider ourselves professionals if we aren't using our tools correctly?

Of course we were all trained in the use of knowledge tools when we went to school, it is just they were last generation ones. We were introduced to paper and pencils and eventually pens. We learnt to use the most basic of information technologies and through the ability to read and write we expanded our memory and improved our thinking.

Since the introduction of mass-produced paper and pens in the mid-19th century, these have been the state of the art when it comes to personal knowledge management tools. But with few exceptions (such as the idea to staple a bunch of pages together in the 1890s, thus creating the first legal pad), there has been limited innovation in this space over that time. Paper is still largely paper and pens are still largely pens.

But with the growth of digital technology, and especially in human scale technologies such as smartphones and tablets, this has changed. We now have access to a range of tools that offer significant advantages over paper and pens. We can carry virtually unlimited amounts of information with us, we can search it to find what we are looking for quickly and easily, we can communicate in real time with other people around the world and we have the flexibility to work effectively from anywhere.

We are moving from a world of specific preparedness to general preparedness. Meetings used to be focused on a set agenda where everyone came specifically prepared to talk about a few key points. If, through the course of the discussion, the agenda changed or new questions emerged then everyone would need to go away, collect all the relevant information and come specifically prepared for a new meeting in a couple of weeks' time. With digital technology, we can now be generally prepared for anything. We can have all our information with us all the time, so if the agenda changes we can immediately access the information we need, and if we don't have the right people in the room we can invite them in over teleconference.

Digital technology is changing what it means to be a professional. It is showing us that even for white-collar work (the type of work that was supposedly immune to automation) the ability to use your tools is as important, or perhaps even more important, than the knowledge you hold. Digital technology has commoditised knowledge and made it increasingly available to anyone who has the right tools and knows how to use them. Creating value as a knowledge worker used to be based on having access to knowledge that other people didn't, but in a world of digital technology where anyone can access information, this is a dwindling value proposition (just as having The Knowledge as a Black Cab driver potentially has less value than having a sat nav and knowing how to use it). Value in knowledge work is about not just accessing information but applying it in a way that improves the lives of our customers.

FROM WHAT TO WHAT IF

Access to information hasn't always been as important as it is today, and in the future you can be fairly sure that it will be more important again. This is partly a result of the developments in digital technology that are improving the quality of information that we have access to and partly to do with the changing nature of work. To understand how the nature of work is changing, we are going to draw a crude distinction between the type of work where you 'do stuff' and the type of work where you 'think about stuff'. The first type of work is 'what' we do, the second type of work is asking 'what if' we do it differently. Although these two types of work are intrinsically linked and we all do at least a little bit of both, their characteristics and how we approach them are fundamentally different.

What we do, or doing work, is the work involved in making and producing things. It is often very procedural in nature, usually relies on specific tools and a certain level of specialisation and, by and large, has a predetermined output. The way we create value through doing work is by increasing the level of output – if we can produce more, we can sell more and we can generate more income.

For much of human history the dominant type of work going was doing work. From the organised hunting of animals, to the construction of the pyramids, to farming and agriculture and through most of the Industrial Revolution, and even in most 20th century offices, the majority of the workforce has been employed to do the work that other people have dictated and in a way that other people have determined.

In contrast, 'thinking work' is much less specific in nature. Rather than being focused on doing it is instead aimed at how to do things better. It is about asking 'what if' and being able to validate or invalidate your hunches. In thinking work there may (or may not) be a clearly defined problem, but rarely a clearly defined outcome. Instead, it is the questioning of what we believe to be true and the exploration of new ideas that lead to the development of new understanding. In a work context, thinking work generally has the objective of finding better ways of doing things. Although there are approaches we can take to improve our thinking work, the outcome of thinking work is more emergent than procedural, and value is not created by doing more thinking but instead by the quality of the outcomes.

Although we have always valued doing work, our relationship with thinking work is a bit more haphazard. There have been eras in human history where too much thinking was inadvisable. Perhaps the most significant example of this is Galileo Galilei. In 1610 Galileo started promoting theories that suggested the earth was not in fact in a fixed position but rather that it rotated and moved around the sun. This was seen to be in conflict with the literal interpretation of the Bible and the response by the Roman Inquisition was to sentence him to indefinite imprisonment in 1633. He died under house arrest nine years later.

So although there has always been a place for thinking work within human society it was not always a safe occupation. In addition, prior to the Industrial Revolution, the domain of thinking work was normally limited to a fairly small and elite subset of society. Few people could afford to spend time on something that couldn't be sold to generate an income. As a result thinking work was left to a handful of church-funded scholars, self-funded inventors and aristocrats who had the luxury of not having to do the doing work.

As a result, before the Industrial Revolution the way things were done was tightly controlled. The guilds, who were associations of tradesman, artisans and merchants, often working under royal decree, tightly controlled the means of production, and the quality and processes associated with a particular profession. In this environment information about processes was closely guarded and innovation was stifled or even actively discouraged. Guilds often exhibited what we would call cartel behaviour and in short, there was little incentive for doing things better and subsequently there was little need for thinking work.

But with the Industrial Revolution came wholesale changes in the way that goods were manufactured and the way the economy operated. Suddenly there was other work to be done apart from the doing work – there was now a legitimate need for thinking work as well. Since then to the present, the amount of thinking work done in society has steadily increased.

In researching his seminal book *The Rise of the Creative Class*, and in subsequent work by both him and the University of Toronto, Richard Florida has observed a significant shift in the composition of the workforce over the last 200-odd years.

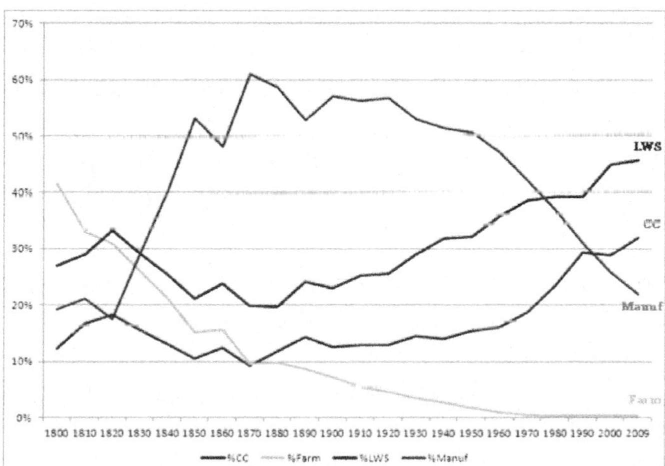

Florida and his colleagues divided the workforce into four distinct classes: Farming and Agriculture, Manufacturing, Low Wage

Services, and the Creative Class (or thinking workers). The above graph shows each class of worker as a percentage of the overall workforce from 1800 to 2009. It clearly shows the rise of manufacturing work as the Industrial Revolution took hold and the subsequent rise in the thinking work of the creative class since the mid-1800s[13].

The data suggests that 30% or more of the workforce now does thinking work, but the real number is much higher than that. Evan Rosen points out in his 2011 *Businessweek* article, 'Every Worker is a Knowledge Worker', that the domain of thinking work is no longer restricted to an elite few – it is now a part of every person's job[14]. When Peter Drucker first coined the term 'knowledge work' in his 1969 book *The Age of Discontinuity*, it was identified as a distinctly different type of work than the more traditional 'manual work'. And although we have seen the rise of the knowledge worker, as shown by the work of Richard Florida, what Evan Rosen points out is that we have also seen the rise of knowledge work amongst manual workers.

Everyone has truly become a knowledge worker, and this evolution has not happened by chance. The rise of knowledge work is a direct result of changes in the operating environment. It is both a result of non-thinking, routine 'doing' work becoming increasingly automated and the rise of digital technology that has allowed the access to information to be increasingly distributed.

While the raw materials of doing work are physical, the raw material of thinking work is information. And just as the development of global supply chains were instrumental in the transformation of doing work, the growth of information infrastructure has led to a transformation in thinking work.

Global supply chains allow iron ore to be mined in Australia or Brazil and shipped to Japan and China to feed the smelters and steel works. This is subsequently turned into white goods that are then distributed to all corners of the planet. Improved infrastructure in both the manufacturing and transportation of goods has had a significant impact. The cost of air transport has reduced by a factor of 10 since World War II and the cost of sea transport has steadily declined over the last 20 years. In addition, goods can be manufactured and delivered much faster than before and this has led to reduced inventory costs, faster product development and many other supply chain improvements[15].

Information infrastructure was once the equivalent of a horse and cart. Actually, scratch that. Information infrastructure used to *be* a horse and cart. Information used to be transported through word of mouth by travellers and merchants as they moved from town to town. But as the perceived value of information increased then specific infrastructure started to develop. In 1815 the Rothschild family famously profited from the end of the Napoleonic wars through superior information infrastructure. By having their own network of agents, shippers and couriers across Europe they were able to get word of Napoleon's defeat at Waterloo a good 24 hours before the government's official messengers. Using this advantage they invested heavily in government bonds and turned them over for a 40% profit only two years later[16].

Just like with the supply chains of physical goods the transportation of information has also become cheaper, faster and more reliable over time. Key to this shift has been the development of modern information technology, or IT. As mentioned above, information is the raw material of thinking work and improvements in IT over the last few decades have shifted the balance towards more thinking work. This is because IT makes distributed decision making and subsequent change not only possible, it also makes it advantageous.

When the operating environment was relatively stable it was possible for the thinking to be done centrally. Information was fed up the hierarchy and used to make decisions, then these decisions were pushed back down again. This was a world where 12 month cycles of strategy, decision making and budgeting seemed adequate.

But as the operating environment for many businesses has become increasingly dynamic, the command and control style of management is no longer fast enough. Just like the shortening of supply chains, the ability for information to move laterally across organisations, societies or even across the globe provides a competitive advantage. We are rapidly moving away from an industrial paradigm where some people do the thinking and other people do the doing. In fact, if all you do at work is doing work then your job is at high risk of automation. The power of digital technology in the distribution of information to front line workers means that now everyone can be engaged in thinking work and it is only this thinking work that will keep us relevant in a world of digital business.

One of the most predictable aspects of technological change is our inability to predict who it will affect most. For much of the Industrial Revolution we have seen muscle being outmuscled by machines. From the first steam engines used to pump water out of mine shafts to computerised robots on automobile production lines, we have become quite accustomed to the displacement of blue-collar work. At the same time we have seen white-collar work as being 'safe' from disruption because the technologies of the time were focused on performing the functions of muscle, not brain matter.

But as Erik Brynjolfsson and Andrew McAfee's point out in their excellent book *The Second Machine Age: Work, Progress, and Prosperity in a Time of Brilliant Technologies,* technology has continued to surprise us in terms of the type of work that it is capable of. There are very few fields of work where people could feel 'safe' – even activities that were considered safe just a decade ago, such as driving a car, have been shown to be at risk. They rightly point out that the safest approach is to "learn to race with the machines", to be able to understand their strengths and their weaknesses and then use them proactively to improve performance.

The best way of determining whether digital technology is relevant to you is to answer the following questions.

1. Are you currently employed (either by someone else or self-employed, either part or full time)?

2. Would you one day like to be employed?

If the answer to either of these questions is 'yes' (or even 'maybe'), then digital technology is relevant to you.

THE GROWTH OF DIGITAL

With all this talk of digital technology, it is worth taking a moment to get some clarity on what 'digital' means. To some, modern technology often appears to be some form of black magic. It allows images and information to fly through thin air, and it recognises our voices and identifies friends and family in photographs more accurately than a police line-up.

Digital is often described as an information technology, or information and communication technology, which are usually abbreviated to IT or ICT. Most people don't dwell on the meaning or purpose of IT because a) it seems incredibly complex and b) most people's experiences of IT are not overly positive and they would prefer to spend as little time thinking about it as possible. If that sounds like you, then I have some good news: I am not here to enthral you with the complexities of IT and how it works; in fact, I am completely unqualified to do so as I have no understanding of this myself. I'm with Arthur C Clarke when he said, "Any sufficiently advanced technology is equivalent to magic."

This may seem like a drawback for someone who shows organisations how to use digital technology, but how it works and how you apply it are two separate things. Most chefs don't know how their knives are made, but they do know how to choose the right knife for the job. So let me tell you just a little bit about what digital technology is and then I can tell you a little more about how it can be useful[iv].

In its broadest sense digital means representing something as digits. As well as numbers, digits can be other discrete letters and symbols. Using this broad definition of 'digital', we can see that digital is not new. Whether it be words hand written on a page, a telegram sent by Morse code or hieroglyphics carved into a stone monument, digital is as old as written language itself.

Digital in all its forms has been incredibly valuable for the capture and sharing of information throughout history. And in the most part it has been something we are quite comfortable with. We are taught it at school when we learn maths, languages or indeed most other subjects.

But over the last 50 or so years the term 'digital' has evolved. Starting with the invention of the transistor in the 1940s, digital has increasingly become synonymous with computing, and we often use these two terms interchangeably. Digital technology is now all around us, not only in our computers; it is the foundation technology of our phones, televisions and radios, and increasingly cars, fridges and even thermostats.

Digital as we now know it uses a binary code, a combination of ones and zeros, to represent other information. For text and

---

[iv] Not knowing how technology works is perfectly normal. In a world of specialisation, people might know how one little part works but no one really knows about the whole thing. For more on this it is worth reading Leonard Read's classic essay "I, Pencil", which points out that not one person has the complete set of skills or knowledge to make something as simple as a pencil. Wikisource, http://en.wikisource.org/wiki/I,_Pencil accessed 12 January 2015

numeric information, this works incredibly well; for more complex information such as images, video and sound it works okay; and for more complex things, such as the representation of experiences in multiple dimensions or the representation of smell and touch, digital does a pretty lousy job still. The reason that digital technology is so useful for information-based work is that once information has been digitised there is no degradation in quality during transfer or duplication. As pointed out by MIT's Neil Gershenfeld, what is truly amazing about digital technology is that it allows us to compute reliably with unreliable devices and communicate reliably through unreliable media[17].

Interestingly, binary code is not a recent invention. Back in 1605 Francis Bacon observed that a binary code could be deployed using any objects at all, "provided those objects be capable of a twofold difference only; as by Bells, by Trumpets, by Lights and Torches, by the report of Muskets, and any instruments of like nature"[18].

This might seem like a nice piece of useless trivia but I think it is important to see that digital is not new – that there is nothing to be scared of, that we have learnt other digital tools in the past and that they have served us well. It is now time to learn some new ones, and these new ones provide an opportunity perhaps far greater than we have ever seen before.

The explosion of personal digital technology over the last couple of decades has its roots back in the 1960s. Back then, most computers were in the hands of government organisations but over time, the power of information technologies has trickled down from governments to organisations, on to project teams and now it is in the hands of individual users. And I mean literally in your hands. Today's smartphones have more computing power than desktops of 10 years ago. Those desktops had more computing power than the supercomputer Deep Blue that famously beat the reigning world chess champion Garry Kasparov in 1997, and Deep Blue had more computing power than all of the computers NASA had in 1969 when they put a man on the moon.

Microsoft describes this shift through the lens of four computing eras[19]. The first era, starting in the 1960s, was the mainframe era and during this era there was one computer for many users. In this era, many organisations didn't have a computer at all and even in larger organisations most people didn't have direct access. This was probably for the best because there was no simple interface for interacting with the computers and you needed to be a specialist technician to have any idea of what you were doing.

The second era, starting in the 1980s, was the personal computing era. In this era, we moved from having one computer for many users to one computer for each user. In this era, we had the ability to access a computer whenever we needed, and as technology improved, so did usability. Up until the late 1980s, you still needed to be able to communicate in computer code to use a computer effectively, but the launch of the Graphic User Interface (using icons and graphics) and the mouse as interaction devices meant that we could all start tapping into the power of technology.

The third era, starting around the turn of the millennium, was the mobility era. In this era, we moved from having one computer per person to multiple computers per person and at the same time computers became both more portable and more powerful. Laptops had started making portability possible during the personal computing era but in reality the size, battery life, connectivity and other limitations meant that most laptops were used like desktop computers. But with the growth of smartphones, tablets and, more recently, smart watches we now have multiple devices to use when away from our main computer. In this era, accessibility has meant moving from having access to a computer when we needed one to having access to a computer all the time.

The fourth computing era that Microsoft describes is yet to come. Starting somewhere around 2020, the ubiquity era will see us have thousands of computers per user. In this era, we will not have to carry a specific device with us as computing power will be embedded into the everyday objects that we use.

This is an extraordinary example of technological advancement but the real story is not the growth in computing power (which is extraordinary), it is the growth in computing access. The problem with computers in the mainframe era was that if someone else was using it, you couldn't use it at the same time. The early supercomputers were so rare that you would have to book time to get access to them and you would need specialist knowledge to use them. This meant that although there were already clear advantages of using digital technology for information processing in the 1960s, most people still had to work in analogue.

This shifted significantly during the personal computing era, however even though access had improved you still needed to be at your desk, plugged into power and connected to the network for it to work effectively. This meant that we still needed to convert analogue to digital when we were at our desks and digital to analogue when we left. The conversion cost of shifting between formats meant that most people continued to use analogue formats for much of their work.

This has only changed in the mobility era, when we have had access to the technology that allows us to capture, store, search, manipulate and share information from anywhere. There is now a clear advantage in using a digital in, digital out approach to work. It is also no coincidence that growth in the power of and access to our digital tools has coincided with the distribution of thinking work in organisations. Digital technology and access to information means that ideas that used to travel at the speed of a horse and cart now travel at the speed of light. In a world of information driven change, the effective use of digital technology is not just desirable, it is required.

BUT IT'S NOT LIKE REAL LIFE

Perhaps the single most common argument against digital technology is that it distracts us and reduces engagement 'in real life'. There is a general perception that the real world is 'better' and that there are aspects of our life that cannot be experienced in other ways. This is partly true, there are many wonderful experiences that cannot be suitably digitised (at least not yet), but there are also situations where a digital representation of real life is good enough, and others where the digital representation is even better.

The computer geek term for putting something IRL (in real life) into the ones and zeros of binary code is 'representation'. The majority of digital information is just a digitised form of real life experiences. We take photos at parties, weddings and other meaningful events, we record and digitise music, we write down or type out the things that were said or that we experienced in our heads. So the argument against digital is not so much that digital isn't real, it is that it isn't real enough.

If you were to see your favourite band play live and had one earbud in, listening to them on an MP3 player you would hear a fairly good representation of the sound coming from the stage. And for most of the time that we want to be listening to music, the MP3 version is good enough, at least good enough for us to willingly trade the quality of a live show for the convenience, cost, reliability and other benefits of digital music.

On the other hand, if we were to add in the full complexity of the IRL music experience, like the smell of stale beer, the body odour, the humidity, the flashing lights and other atmospheric aspects that we experience in the three dimensions of a live music venue, we see that MP3 offers a very poor representation.

In reality, we don't always want to experience music in this environment. I don't get to do it in the comfort of my own home, and I can only do it when and where the act decides they want to play. There are also some very real costs of attending live music – there is the cost of the ticket, the time and cost associated with travel to the venue, the overinflated prices of refreshments, and I have to share the experience with a bunch of other people that I don't know. Sometimes we want to have some chilled out background music at a BBQ with friends or use music to get us pumped for a workout or a run. In such situations, the digitally represented version of the live music experience is not only good enough, it may actually be better.

On the other hand if you were to compare the quality of your interactions with your friends on Facebook versus IRL you would probably find that the ability to convey your thoughts and feelings is significantly hampered by technology. Depending on the complexity of the message being shared research shows that as little as 7% of the message is communicated in words and the majority is through body language and tone of voice[20]. Digital platforms often can't provide the level of resolution to convey the full complexity of human interactions and our messages become pixelated. This is also why we shouldn't have important, emotive or other difficult conversations over email or via text message, as these channels provide too much scope for miscommunication.

In the 21st century this is an incredibly important skill to develop – knowing when to choose the real life experience and when the digital representation is good enough (or better). What is the rich experience going to give us, and is it worth the cost? It seems like this trade-off is not necessarily a static one; over time we have become more accepting of the digitally represented alternative. This is partly because over time we uncover additional benefits to the digital version that weren't obvious when we thought it was a pure substitution, and these additional benefits make the trade-off worthwhile.

Take for instance the electronic calendar. Initially we thought we were just trading our perfectly good paper diary for a somewhat strange looking digital one. What it turned out we were getting was a super diary that allowed you to sync your appointments across all your devices, create reoccurring appointments, edit appointments with a few clicks, check people's availability for meetings and keep track of who has confirmed. But nobody knew all of this until after they started using it.

In many ways digital has become the new normal, and our brains have even started rewiring themselves in a way that makes the new normal seem more like the old one. A good example of this is research recently conducted by Flinders University showing that humans now respond to emoticons, the small faces that we create using our keyboard, in a way that approximates our response to human faces. If we look at a ':-)' or ☺ we get an experience similar to what we would get if someone looked at us with a smile on their face. In some ways we could consider emoticons to be highly pixelated faces[v].

The research is quite limited at the moment, but it shows that as humans, we have developed a recognition of a smiley emoticon as

---

[v] Perhaps somewhat ironically it is not just people who have started to see the combination of a colon, a dash and a close bracket as a face. Many of our digital tools now recognise it as a combination of symbols representing a smiling face and autocorrect it for you ☺.

a symbol representing happiness. It's not entirely surprising given that we respond to symbols which approximate feelings and emotions all the time. For example, if we look at the word 'love' as it is written on a piece of paper, it is just a bunch of symbols. Through learning and association, we associate particular meaning with that set of symbols, and the written word 'love' becomes a pixelated version of the real life experience of love.

Over time we have come to associate strong meaning with different combinations of alphanumeric symbols. In the hands of a great writer or poet, these symbols can elicit feelings and responses of great intensity. In fact, we no longer see these symbols as an approximation of the physical experience but an experience in their own right, with its own unique characteristics that the physical experience might lack.

For example, a book can capture a moment and store it for eternity. It can be shared and reflected upon by thousands or millions of people regardless of distance and time. I would say the very best books are the ones that are least like our real lives, as they are the ones that we reread and get to experience over and over again, whereas real life moments happen just once.

But just like with the story of the MP3 player, there will come a time when we're not willing to accept the approximation in place of the rich, real experience anymore. Increasingly we are starting to see the challenges of engaging in real life events as positive aspects of the experience. On the one hand, I see this as a desire to still make commitments in a world when so many things have become transient and easy. On the other hand, I believe that it is tied to our need for human contact, something that is innately human.

Largely we have been more willing to accept the pixelated version of IRL experiences in business than we have in our personal lives. We have books, shelves, filing cabinets, email inboxes, hard drives and servers full of files, documents and information that represent

IRL experiences. We are forever writing reports that provide a documented representation of a past experience or future opportunity, we write meeting notes that are a representation of what was said and implied in a real life conversation and we create budgets and financial reports filled with numbers that are nothing more than a (often highly manipulated) representation of real life value.

We have become acutely aware that such representations can be manipulated, that words can be taken out of context and that financial statements can be used to deceive. In a sense this is no different than the risks associated with putting too much faith in digital technology, of not knowing its limitations and believing every pixel in front of us. We shouldn't believe everything we read on Wikipedia any more than we should believe everything we read in a printed newspaper.

As the power of digital technology continues to grow, our ability to represent real life will become more accurate and more lifelike than ever before. With this comes great opportunity – we will be able to make better, more informed decisions and more accurately meet our customers' needs. The growth of digital technology in business is testament to the value creating opportunity that it offers. But as the accuracy of our representations grows, there is also a risk of being seduced by it. We may place too much faith in the certainty of absolute numbers, words, language and pixels in what is inherently a complex and uncertain world. As we move into an era of digital business, it is just as important as ever to know that it is not like real life.

YOU GET PAID FOR EUREKA

We probably all know the story of eureka. I remember reading the kids' picture book when I was in primary school, complete with illustrations of a naked Archimedes running down the street. The story goes that Archimedes was tasked with solving a problem for his king, Hiero of Syracuse. Hiero was pretty sure that when minting coins his goldsmiths were removing gold and replacing it with a lower value metal. Although he knew the weight of the coins, their irregular shape meant that he couldn't accurately determine their volume and subsequently their density (an important indicator of purity). In the story, Archimedes was thinking about the problem in the bath and he realised that when he lowered his body into the bath the amount of water displaced was equal to the volume of his submerged body. He was so excited by his discovery that he jumped out of the bath and ran down the street exclaiming "Eureka!" meaning "I've found it".

Unfortunately, the story of Archimedes is somewhat deceptive as it has led most of us to associate innovation with world-shattering breakthroughs created by boffins, scientists and Silicon Valley startups. We imagine it as a creative exercise done in isolation and ending with a flash.

We see eureka like winning the lottery of ideas, something that we stumble on because of a small amount of preparation and a large amount of luck. But in the words of Thomas Edison, "I never once made a discovery. All my work was deductive, and the results I achieved were those of invention, pure and simple."

Eureka is much more about uncovering everyday opportunities and answers to everyday problems. And those eureka moments, those ideas, no matter how large or small, are the outcome of a process. The process of thinking work is based on asking 'what if', examining the evidence, and creating responses that have the potential to offer positive and valuable change for our customers. This is what thinking workers get paid for.

This is very different from the value created in doing work. Value in doing work is about production volume. In the short term if you have one production line that delivers a profitable product then the best thing you could do is have 10 lines producing the same item. Assuming that you don't satisfy all the demand, your profits will multiply. But in the long term conditions will be different, competitors will replicate or improve on your product, consumer tastes will change, input costs may change. A profitable product will generally become less profitable over time, and may eventually be produced at a loss if no changes are made to update the product or the process. And if you had decided to replicate your production line to maximise profits in the short term, in the long term this will multiply your losses.

The only way to make profits in the long term is to constantly create change and improve on the status quo. It is this focus on questioning and ideation that, at least for now, separates people from technology. Good decision making in a complex world requires more than cognitive ability, it requires the moment of creativity and ideation. This is the moment of human glory, the bit that technology can't do for us, the bit where we come up with new ideas. But as Matt Ridley points out in his outstanding 2010 TED Talk "When Ideas have Sex"[21], our ideas are not new – new ideas are a recombination of old ideas in a new way. And the way that we get access to old ideas is through information exchange.

Matt suggests that it is this ability to exchange ideas and effort that is perhaps mankind's greatest invention[22].

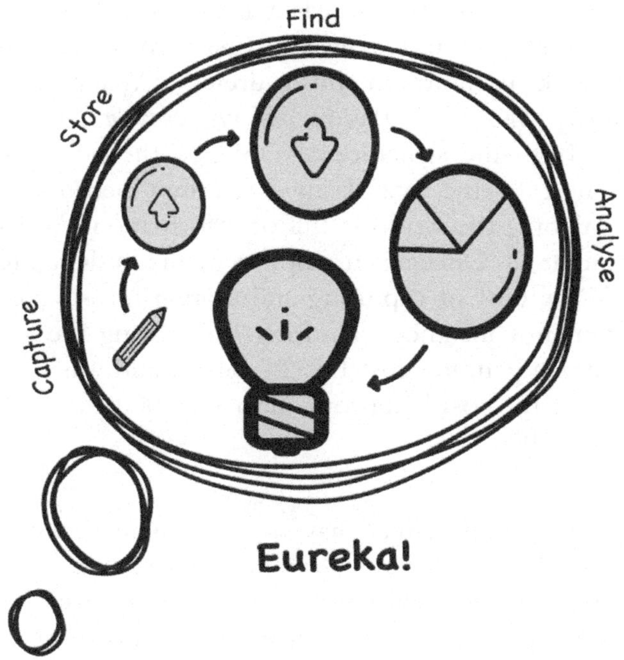

Although that moment of creating the new combination of old ideas into a new idea seems innately human, there is a process to ideation that can be aided by our digital tools. Before our eureka moment we need to be able to analyse information and other ideas available to us and it is almost impossible to comprehend how difficult this might be in today's world without the use of digital technology. The ability to compile and manipulate information has made ideation faster and more effective than ever before.

Taking a step further back, before we can analyse information we need to find it. And once again the ability to find things quickly through searching helps us get to eureka faster. Do you need to know how pivot tables work in Excel? Search YouTube. Want to know the location of all outdoor picnic tables on the Gold Coast?

Go to data.gov.au. Need to know how big your customer's roof is so you can quote that next solar installation? Use Google Maps[vi].

Finding all this information so easily is fantastic, but of course for us to be able to find and analyse it, first it must be captured and stored. As Erik Brynjolfsson and Andrew McAfee point out in *The Second Machine Age: Work, Progress and Prosperity in a Time of Brilliant Technologies*, from the structure of DNA to the position of roads and buildings, to the availability of a two bedroom house in Phoenix Arizona, the world is full of information. And slowly we are digitising it all. Once again, improvements in digital technology have made the cost of capturing and storing information cheaper and cheaper. For instance, capturing and storing the sequence of the first human genome cost US$2.7 billion and was completed in 2000. By 2011 the cost had dropped to $100,000 and in 2015 it will fall to about $100.

In this process we effectively have a value chain. A value chain, as described in Michael Porter's 1985 book *Competitive Advantage*, is a sequence of activities that a firm undertakes in order to deliver a valuable service or product to market. In this case we have a sequence of activities starting with the recording of the information and ending with the communication of advice. From recording, the information is stored, at some later point often separated by both geography and time, and we then search for and find that information. After finding it, the information is analysed and we create a response. Finally, after all that we provide our advice to the customer, and hope they like it.

[vi] Sites such as http://www.freemaptools.com/ have an area calculator function that allows you to calculate the area of an irregular shape on Google Maps.

As much as we would like to think that all the work we do is valuable, there is only one point in this chain where value is created, and that is the point where we provide the product to our customer. Every other part of that process is actually a cost. It costs money to record information; it costs money to store it; and it costs money and time and resources to search for, find and analyse it. It even costs us time and money to be creative and generate a response. The only time that we generate value and get paid is when we deliver the output of our work, whether that be advice or making a decision. If we start to think about our information management as being this flow from the recording through to the advice, then we start to think differently about the way that we manage our information.

To look at the value chain more closely, the first thing we need to do is separate the individual processes from the outcomes. The *process* of recording or storing information is not itself valuable; it is the outcome – the record or the stored information that generates the value. A prime example of this is when we have not filed something correctly and we spend hours looking for it, it doesn't make the information more valuable than if we had put our hands on it straight away. What we want to do is spend less time looking and more time finding.

The time spent looking for information comes at a cost; if we spend more time looking for things, that is less time we can spend finding things, creating things, or responding to our customers' needs. This is one of the massive advantages of digital technology. It makes the finding of information faster so that we end up spending less time looking.

The other great advantage of working with information in this way is that the generally accepted format for sharing information with others is now digital and we can save ourselves time converting information from analogue to digital and back again. It is expected that when we send a report to a client, the main version is electronic, or at least that if a hard copy is required (most likely printed from the electronic version anyway), it will be accompanied by the electronic version.

So if the standard format for communicating information is digital and we need to put our information into a digital format at some point anyway, it makes sense to record digitally in the first place. One of the biggest time savings of going digital is avoiding the rework associated with doing something once on paper and then retyping it into an email, proposal or CRM system. We need a digital in, digital out approach to our work.

Digital technology enables us to access and recombine an almost infinite number of past ideas into new ones, test those ideas faster, and measure the results more accurately than ever before. In the business of thinking work digital technology can make your work both easier and better.

And that is all that eureka is about. Rather than going to work to do your job, go to work and do your job better. Regardless of your job title or the industry you are in, we all get paid for eureka.

WILL YOU BE RELEVANT?

So here we are on the cusp of a revolution, as significant as the shift from the Stone Age to the Bronze Age but happening a whole lot faster. The speed of modern business means we aren't going to have thousands or even hundreds of years to respond – we will be lucky to get another 10. This means we need to make decisions now about the digital revolution. The impact is already being felt. Digital is already shaping the corporate landscape; it is infiltrating our workplace and almost no job is immune.

There is a good chance that your job as you currently know it will not exist in a decade or so. It may evolve into something that is still familiar but just as likely it will be replaced. As Andrew McAfee and Erik Brynjolfsson point out in *The Second Machine Age*, a 'computer' used to be the job title of people who did manual calculations, and now it is a machine.

For much of the time since the Industrial Revolution, business has been obsessed with specialisation as a way of improving business efficiency. This means getting workers to focus on a narrower and narrower set of expertise, and a narrower and narrower set of tasks. For some of these people their job description might will end up being the tasks that technology does, and for those people it is unlikely they will be able to compete.

For others their displacement by technology will be less clear-cut. Instead of being replaced by a robot or a computer program, technology will just slowly chip away at their value. For them it will be a death by a thousand cuts, with each new technology slowly undermining the value proposition of their employment until one day someone does the maths and decides that their compensation is greater than their value, and they are let go.

But the fall of our current working arrangements will give rise to something else. Just as the Industrial Revolution saw the rise of new types of workers, including much of what we now call white-collar work, the Digital Revolution will also provide new opportunities and these opportunities will present themselves in every job, in every business and in every industry.

In the book *Exponential Organizations: Why New Organizations are Ten Times Better, Faster and Cheaper Than Yours (And What To Do About It)*, Salim Ismail opens by talking about Iridium moments[vii], moments where organisations over-invest in the present and under-invest in the future because they fail to update business assumptions, specifically those relating to digital technology. Salim points out that there are many other organisations such as Eastman Kodak that have also had their Iridium moment, and given the accelerating performance of digital technology there are going to be many more to come. But it is not just organisations that have to deal with Iridium moments – they are just as relevant to individuals. If we don't update our assumptions, if we continue to

---

[vii] An 'Iridium moment' occurs when linear tools and the trends of the past are used to predict an accelerating future. It is named after the company Iridium that in the 1980s launched a number of low orbit satellites to provide mobile phone coverage in places where mobile phone towers were not economical. Iridium failed spectacularly and cost investors more than US$5 billion dollars. The cause was the improved economics of mobile phone towers as network performance and other digital infrastructure improved in performance and reduced in cost making the cost of satellite telephone services uncompetitive.

try to protect ourselves against emerging technology, we will not win. We cannot outcompete what is fundamentally an evolution in the way we do our work.

In his book, Salim shares a quote from David S Rose, author of the bestselling book *Angel Investing: The Gust Guide to Making Money and Having Fun Investing in Startups*, that sums this up perfectly: "Any company designed for success in the 20th century is doomed to failure in the 21st." And any worker optimised for success in a world of analogue business is doomed to fail in a world of digital business.

If you want to ensure your future relevance then you could do worse than take the advice of Professor David Tuffley, lecturer in applied ethics and socio-technical studies at Griffith University. In a recent article for the *Washington Post*[23] titled 'In 10 years, your job might not exist. Here's how to make sure you're still employable,' he points out that firstly we need to continue to work on our thinking skills as it is the ability to analyse information and recombine ideas in new ways that provides a competitive synergy with our digital tools. Secondly, we need to be able to use our technology and be comfortable with it, we need to master new media, manage our information effectively and be comfortable working in virtual environments.

Finally, he advocates the importance of continual growth and diversification in our skills and knowledge. It is incredibly difficult to predict which industries, which jobs and which tasks technology is going to affect next. For a long time white-collar work was considered too complex for digital technology to undertake, but even the once-safe legal and accounting professions are now under threat. There are now computer algorithms undertaking the work of lawyers doing document discovery, and the procedural work of bookkeepers and accountants is being replaced by automated information feeds and machine learning. The only way to ensure your relevance is to hedge your bets and ensure you have a diversity of knowledge and ability.

But just because it is unavoidable that doesn't mean it is all bad. It turns out that digital technology is very effective at doing some of the tasks that we hate the most. There is a massive opportunity to spend our time doing more enjoyable, higher value work. These are the rewards on offer in a world of digital business and the next section is all about that.

REWARDING

TECHNOLOGY MIGHT MAKE YOU WORTH LESS BUT NOT WORTHLESS

After reading the previous section I hope you aren't feeling too disheartened. It is true that the robots are coming, but that isn't as bad as it seems. In every challenge that this change brings, there is an opportunity. The biggest opportunity is in doing something the robots can't, and using the robots as a tool to do the uniquely human aspects of your job better.

Let's be clear about what I mean by robot. A robot is simply 'a machine capable of carrying out a complex series of actions automatically'[24], and it can be either a machine undertaking physical tasks (a la HAL 9000 from Arthur C. Clarke's *2001: A Space Odyssey*) or a computer program or computer algorithm undertaking an informational task.

Most robots are used for a single function. Robots like tasks that consist of simple, discrete actions that need to be repeated over and over again. As Erik Brynjolfsson and Andrew McAfee point out in their book *The Second Machine Age*:

> *When it comes to work in the physical world, humans also have a huge flexibility advantage over machines. Automating a single activity, like soldering a wire onto a circuit board or fastening two parts together with screws,*

is pretty easy, but that task must remain constant over time and take place in a 'regular' environment.

So although the robots are coming, in the short to medium term it is incredibly unlikely that one of them will take your job. Rather than replacing entire jobs, robots replace individual tasks. In most cases jobs are complex and involve a number of objectives made up of different activities, which in turn are made up of a number of tasks. On top of that there is the complexity of interpersonal relationships, creating meaning from spoken and written conversation, identifying underlying needs and exercising judgement in decision making.

It is this complexity that makes jobs interesting for people, and incredibly challenging for robots. Programming a robot to manage the complexity of your whole job would be incredibly difficult, if not impossible (this may not always be the case, as computing advances will continue to make the impossible possible, but for the immediate and near future this will hold true).

Another point of reassurance is that humans have a flexibility advantage that allows us to adapt to different tasks and situations quickly and easily, though technology is often cheaper over time. This means there are different economic advantages of both people and technology. Technology generally has a high capital cost associated with development and setup but a low operating cost – whether it be a steam driven pump to pump water out of a mine at the beginning of the Industrial Revolution, or a modern computer algorithm to produce relevant web search results, a large amount of time and money goes into purchasing the technology, setting it up to work correctly and testing it to make sure that it works as advertised. Once operational, technology starts to repay the initial investment by repeatedly doing the same process at a marginal cost lower than the task's value. A financial return on technology is only generated over the medium to long term once the total capital cost has been recovered. What this means is that the economics of technology is fundamentally about volume and scale.

People, on the other hand, are quite the opposite. Relative to technology people have a low capital and setup cost, but much higher operating costs. For a one-off event, a person might be cheaper – sending a miner down the mine once or twice with a bucket, or getting an individual to scour web pages to find ones relevant to you is going to cost less overall than purchasing new technology. But when you want to repeat the same action over and over again, the high operating cost of people means that humans become relatively expensive. Whereas the economics of technology is in the scale, the economics of people is in diversity and flexibility.

Because technology requires scale to drive value, perhaps the defining characteristic of the work that technology does well is that it is common. A good example of this is searching for information on Google. Google handles about 3.5 billion searches per day; about 15% or about 500 million of these are unique search terms that have never been seen before and the remainder are search terms that Google has previously handled. But regardless of the search term the *process* of searching Google is exactly the same: you type your search term in the text box or web browser and Google delivers responses in a standardised format and layout, with ads at the top and down the sides and your search results in the middle.

Even the process that happens behind the scenes is standardised and automated. The results you receive are dependent on an algorithm (which is nothing more than a set of automated processes) that takes into account over 200 variables including keywords found on the website, the freshness of the content, and your location, but the process itself is still consistent.

This type of scale means that the setup and infrastructure costs associated with Google's search business are spread across trillions of searches each year. So even though this infrastructure might cost billions of dollars to run, the cost of each search in real terms is only a fraction of a cent. As long as users click on a paid ad every once in a while Google can afford to offer its search service for free, which means we get to search for whimsical things such as 'how animals eat their food' on YouTube (go on, I dare you to try it).

The level of commonality is therefore an incredibly important aspect of the economics of a task. If a task is incredibly common then it doesn't have to be worth much before it gets automated. Technology is of course capable of carrying out low volume tasks, but the lower the volume the higher the value needs to be before automating it is worth it. Over the last few years we have started to see automation creeping into labour intensive areas of the legal profession, especially in relatively common and information intensive tasks such as document discovery. Although a relatively specialised service, the economics stack up because it replaces junior lawyers that might otherwise be charged out at hundreds of dollars per hour.

The final factor in deciding what gets automated, in addition to the commonality of the process and the value of the task, is stability. If a task constantly changes then whatever technology you employ to do it will need to be constantly updated. Rather than a capital cost like the initial setup, updates are more like an 'operating cost' to keep your technology running. But like setup costs there are different economics for people and technology. People are generally quite flexible, and creating small variations in their activities is neither time consuming or difficult. On the other hand, creating these same small variations for technology could require both a recalibration and another phase of testing and commissioning.

In summary, technology is generally suited to common, stable, scalable activities whereas people are more suited to the dynamic and the unique. Technology does ordinary and people do extraordinary.

There is much debate about what digital technology is and isn't capable of doing but in reality digital tools are created by people who are guided by the same economic considerations that you are. It is highly unlikely that any right-minded investor or programmer is going to back the development of a technology to undertake tasks where the operational savings are too low or the maintenance or capital investment is too high.

And this is good news for people. Our jobs are full of procedural tasks that are in the process of being automated – the simple, repetitive tasks that need to be done over and over again with little change over time, tasks that require precision but no creativity, tasks that require you to follow orders and not ask questions. If you think about it, the kinds of tasks we generally don't like that much. We also all have a set of skills and innate human abilities that technology cannot replace and cannot replicate (at least for now). Technology doesn't know what we want, it cannot form meaningful relationships and cannot think creatively about the future. These are the skills that make us human, that we generally enjoy the most and that will define our value over the long term.

So while technology has the potential to make us *worth less*, it is unlikely that we will ever become *worthless* and in fact there is ample opportunity for technology to make us worth more. This is not to say that technological unemployment doesn't occur but it generally is not the case of a robot directly replacing a human being; it happens because the economics of employing that person diminish over time. Parts of a job get automated bit by bit and at some point the cost of employing that person exceeds the value of the tasks they complete, and then through a process of rationalisation three jobs get rolled into one and technology does the rest[viii].

Or more likely this same economic redundancy will happen on a broader scale. It won't be the individual that becomes uncompetitive, it will be the whole organisation. Throughout history organisations have evolved more slowly than the marketplace as a whole. New competitors who use better technology and improve processes will emerge. Older organisations will keep doing what they've always done, and as more productive competitors reduce prices, profit margins will slip. One day the cost of doing business will become greater than the income and the business will fold[ix]. People will then seek re-employment or up-skilling in other organisations that are more profitable (and likely to be more automated and technologically advanced).

There is a saying in business, 'Do what you've always done, get what you've always got'. It is a piece of advice that suggests that you can't get better if you aren't willing to change. The flip side of this advice is that if you are happy with what you are currently getting then there is no need to change. But as Yvette Adams,

---

[viii] Sometimes, in the name of productivity organisations pursue a strategy of specialisation, and create jobs that are effectively the same task done over and over again (like the person who flips burgers at your local fast food joint, the data entry clerk at the accounting firm or the person who used to tighten the wheel nuts on the cars before they rolled of the production line). These hyper-specialised jobs generally face the highest risk of automation because a single purpose robot can be created to carry them out. If this sounds like your job the best thing you can do is to build as much diversification into what you do as possible.

[ix] For a much more detailed explanation of this I highly recommend *The Origin of Wealth* by Eric Beinhocker.

author of *No Kidding: Why Our Kids Know More about Technology than Us and What We Can Do about It* once pointed out to me, this is blatantly untrue. These days if you do what you've always done you no longer get what you always got. You are slowly but surely getting less. We live in a world of accelerating change and a choice to maintain the status quo is a choice to become worth less.

Rather than focus on the loss of low value, repetitive and menial tasks – which, if we are being honest, are not the tasks that most of us want to do anyway – we can embrace the opportunity to offload all of these tasks and spend more time on the parts of our jobs that we do like. Of course, a certain amount of change is required (and we will talk about that in the next section), but the outcome of that change is the opportunity to focus on the work that is more creative, more valuable and ultimately more human.

THE DIGITAL ADVANTAGE – ACCURATE, FAST AND CHEAP

For more than 35 years up until 1996, Guatemala was in a state of civil war. During this time it is estimated that more than 200,000 people were kidnapped, tortured or killed, or otherwise went missing in state sponsored activities. Most of this was done in secrecy with little evidence of what had happened. Then in 2005, the random inspection of buildings near a Guatemalan police compound led to the discovery of a massive archive of information. With millions of documents in various states of decay (some of them dating back to the 1800s), the archive is believed to contain answers for many of the missing people's families.

In total, the archive is estimated to contain more than 80 million sheets of information and with funding from a number of European countries a team of experts has been working to digitise the archive. The aims of this are to preserve it for future reference (there has been at least one attempt to firebomb the archives) and aid in understanding what transpired during the course of Guatemala's civil war. So far the project has resulted in the sorting and digitisation of about 15 million documents, but they estimate it will be another 15 years before all the documents have been processed.

So why digitise the information in the archive? What are the advantages of digital technology over analogue that make this 15 year project worthwhile? There are innumerable advantages, but most of them can be tied back to three key characteristics: accuracy, speed and price.

Accuracy is a strong advantage of digital technology. As discussed earlier, digital technology has always been incredibly good at representing information such as text and numbers and is now becoming increasingly good at representing more complex forms of information, such as sounds and images. What is more, exact copies of this information can be produced with no degradation in quality or accuracy.

Digital technology is now reaching a point where not just text and numbers but also images, video and sound files can be searched accurately as well. The music recognition service Shazam allows you to identify a song by holding your smartphone or tablet microphone up close to a music source. Its app generates an 'acoustic fingerprint' of the track you are listening to and searches a database of approximately 11 million songs. If it finds a match it delivers the details of the artist, song title and album back to your device. In October 2014 the developers announced that Shazam had been used to identify more than 15 billion songs.

Another example is Google's image recognition algorithm that can now correctly identify the content of an image with greater than 90% accuracy. A recent test of the algorithm using a database of more than 14 million images highlighted its accuracy. It was not only able to distinguish between different species such as cats and horses but it was also able to distinguish between more than 200 different breeds of dog.

Compare this to the quality of the analogue information captured in the Guatemalan Police Archives – hand written notes, typed reports, bad photocopies and grainy images – and you could imagine how much easier it would be to uncover relevant information if it was contained in digital records. Accuracy makes information easier to find, easier to share and easier to manipulate.

The second characteristic of digital that provides a distinct advantage is speed. Information can be repurposed and shared incredibly quickly. Creating the first digital copy of the Guatemalan National Police Archives is likely to take 20 years or more, while the second digital copy will be ready in less than a minute. The same speed advantages also apply to search functions. Google will search its archives of web content and deliver you about 14,000 relevant search results for 'Guatemalan police archives' in less than half a second, but the search for a single name in the original analogue archives, even with a team of experts working around the clock, would be an almost insurmountable task.

The final characteristic of digital to consider is cost. Digital technology has made information storage incredibly cheap – in many cases the cost is approaching zero. Although commercial rental rates in Guatemala are quite reasonable, 80 million sheets of information still take up a lot of room. In its current form the collection fills nearly a dozen rooms, many of them stacked floor to ceiling. If the information were to be sorted and organised into filing cabinets it would fill nearly 700 three-drawer filing cabinets. Digitised, this same information would fit on one 512 GB USB flash drive the size of your pinkie finger, and the cost of the flash drive would be similar to buying one filing cabinet (other digital storage options such as traditional hard drives and cloud storage would be even cheaper).

Digital information is also incredibly cheap to reproduce and share in comparison to analogue information. The only real cost in reproduction is the purchase of the storage medium. In comparison, to reproduce the original archive using the self-service photocopier at your local office supply store is likely to cost you around $6.4 million dollars, take about two years to complete (assuming you could find some mates to help you out with working a 24 hour shift) and produce a copy of lower quality than the original.

The advantages of digital are already significant but digital continues to become more accurate, faster and cheaper all the time. Computer processor speeds are still improving, search algorithms are becoming more accurate and the cost of online storage, the thing we often refer to as 'the cloud', is now approaching a point where it is often thrown in free like a set of proverbial steak knives. It is reasonable to ask why you would bother creating and storing new information in a format *other* than digital.

In terms of the commercial advantage of digital technology, it is worth noting the relationship between these three characteristics and business profits. There are only three fundamental drivers of business profits, which align directly with the three advantages of digital. The first is the price you can sell your goods and services for, and improved accuracy impacts quality, and subsequently the price you can charge. The second is the volume you can sell; speed improves productivity which directly impacts the volume that can be produced. And the third is the cost of production, the sum of money required to produce each item sold.

The drivers of business profits are just as relevant to the individual. Your 'profit', or what you get paid, is derived from the quality of what you do, the amount of work you can complete and the costs associated with you doing the work (this might include rent, IT, administration and other overhead costs). If you can do more high quality work at a lower cost you are ultimately worth more. We don't often look at our work like this because we only see the job requirements and a wage but over time quality, output and costs will all play a part in our future employability and remuneration. Therefore the use of digital tools to improve both efficiency and effectiveness will positively impact our own commercial success as much as it impacts the commercial success of our organisation.

MULTIPLYING RETURNS

When we look at replacing an analogue process with a digital one, we tend to think of them as different versions of the same thing. For instance, if we were to 'take notes' digitally on our tablet, we think of it as an alternative method of taking notes on paper. The process would look much the same as taking notes on paper and both processes would produce a set of notes.

To help us make replacements of this kind, many app and software developers have applied a design principle called skeuomorphism. Skeuomorphism means making one thing look like another thing that it is not. That might make something appear more valuable, such as a fake wood grain panel in your car actually made of plastic, or increase familiarity, such as a writing app that uses a yellow striped background so it looks like a legal pad.

In some ways skeuomorphism is really useful, as it can make unfamiliar digital technology a little easier to engage with by providing recognisable visual cues. These cues can help us to quickly learn how our new digital tools work and help new users understand what an app is meant to be used for. As a result there are many app icons that imitate calendars, notebooks, envelopes, maps, clocks, task lists and other real world equivalents.

When we create a direct comparison between digital and analogue processes, the downside is that our understanding of one can create limiting beliefs about the other. For instance, if I 'know' that the fastest way to take notes on paper is using a pen then I would most likely believe that the best way of taking notes on my tablet is with a stylus (the digital equivalent of a pen). But this belief is limiting. It turns out that we can take much faster and higher quality notes by typing them rather than writing them. In addition, digital notes might also let us link to other notes, keep different versions, record our voice, or link to external sources like webpages or documents. If we are focussed on using our digital note taking as a basic substitution for hand written notes, we may miss out on the expanded opportunities the new tool offers.

Another problem with skeuomorphism in digital technology is that it might stop people engaging with the tool in the first place. If you believed that taking notes digitally is just like taking notes on paper, there would be little incentive for you to engage. What would be the point of investing in an expensive digital device and spending precious time learning how to use it (in the process potentially putting your professional reputation at risk in front of clients and colleagues as you fumble along) when there is no perceivable advantage?

Unfortunately, this is often the main barrier that stops people from engaging with new digital tools. For these people, to understand what digital can be used for, they need to know what the tool is comparable to, but as soon as they know what it is comparable to then they can no longer perceive the differences that drive benefit.

It is only through engagement and use that the additional benefits of digital technology reveal themselves. Initially these benefits are largely in the form of productivity gains. By working in digital we can capture information more easily, store and retrieve it quickly and share it with a touch of a button. In comparison, our analogue systems are incredibly ineffective. The fastest we can capture written information is about 30 words a minute, about half the speed of a proficient typist, and what is captured on paper varies greatly in terms of legibility and is often not fit for anyone's consumption except the author's. Once captured it is difficult and time consuming to store and search, usually involving some combination of lever arch folders, filing cabinets or dusty archive boxes and offsite storage. Then when it finally comes to sharing the information with others or reusing it the whole thing would most likely need to be redone digitally anyway.

The exact returns generated from productivity gains can be difficult to calculate and depend on the work you do and the way you do it, but research by Google suggests that people engaged in knowledge work spend more than a third of their time looking for information. Data collected from individuals who shift from an analogue to a digital centric approach achieve productivity gains of between 10-20% in the first year.

But the returns of digital technology are ultimately much bigger than the productivity gains. As discussed earlier, one of the fundamental differences between doing work and thinking work is how value is generated. Whereas productivity gains have the potential to provide additive benefit through a one-off increase in output, modifying the way we do things, the outcome of our thinking work can have a multiplying effect.

New ideas are expensive to come up with but cheap to share and reuse. Carl Shapiro and Hal Varian first alluded to this idea in their 1998 book *Information Rules: A Strategic Guide to the Network Economy.* In the book they state, "Information is costly to produce but cheap to reproduce", and give an example of the creation and distribution of a blockbuster movie. The budget for production may exceed $100 million dollars, but to reproduce a copy on DVD or to distribute over the internet might only cost a few cents.

So in terms of thinking work, an original idea that leads to a 10% reduction in packaging costs might be costly to develop but it can be reapplied relatively cheaply in similar situations – this is sometimes called 'best practice'. Every time the idea is applied successfully it has the potential to generate an additional 10% reduction in packaging costs. If it is implemented 10 times you will get a return approximately 10 times as large as if you implemented the idea once.

The real power of digital technology in this process is that it can provide opportunities to come up with better ideas more often. As we talked about in the previous section, most new ideas are just the recombination of older ones, and our digital tools have changed the way that we access, analyse, and even create new ideas.

In analogue business accessing new ideas was time consuming and difficult. You had to attend trade shows and conferences, read trade publications (available when the publishers decided to publish the information rather than when you needed it) and travel to meet people face to face. In this environment most people were excluded from sharing their ideas, unless they were an established industry professional or academic.

But in a world of digital business, most new ideas are digitised for distribution and increasingly made available over the web. Nearly everyone with internet access now has the ability to publish their ideas openly for other people to see, share and build on. In his excellent book *Smarter than You Think – How Technology Is Changing Our Lives for the Better*, Clive Thompson discusses how the internet and digital technology are making us smarter. Firstly, digital tools have given us prodigious external memory and we have greater ability than ever before to record, store and recall our previous experiences. Secondly, the ability to easily link knowledge, people and ideas makes it easier than ever before to join the dots between disparate pieces of information. Finally, digital platforms have created a space for 'thinking out loud' and the quick and easy exchange of ideas across geography and time.

Digital technology allows us to create new ideas faster, and by modifying the way that we interact with information it also allows us to generate better ideas than we could previously. And once a new, valuable idea has been generated it is likely to be encoded in the standard format for today's business, digital, for distribution. It could be a report or email but could just as well be a video, blog post or eBook. These ideas then join the multitude of other ideas flowing through the Internet and seed the next generation of ideas to come.

PERFORMANCE ENHANCING TECHNOLOGY

In *Smarter than You Think*, Clive Thompson recounts the story of chess master Christoph Natsidis, who was caught using a chess program installed on his mobile phone during the 2011 German Chess Championships. During the game, he kept vanishing to the bathroom for up to 10 minutes at a time. His opponent became suspicious and confronted him, along with the referee. When they found his smartphone with the game's moves programmed into it, he admitted to cheating.

This is not the first example of a player using a computer to compete at chess, but what was different in this case was that Natsidis was already very good. During the tournament, he had already achieved the rating of International Master, only one rank below Grandmaster. As Clive Thompson points out, "Chess had entered a phase similar to the doping scandals that have plagued baseball and cycling, except in this case the drug was software and its effect cognitive."

In any competitive environment, we expect professionals to push the boundaries of their physical, mental and technological capabilities. You won't find a professional tennis player on the ATP tour using a wooden racquet anymore because, as former world number one Björn Borg showed in the late 1980s, it is almost impossible to be competitive using old technology when everyone else is using something newer and better.

We may look back with nostalgia to the days when the likes of Rod Laver, Ken Rosewall and Margaret Court moved gracefully across the court, artfully dropping the ball over the net, but the game has moved on. Developments in racquet technology have created racquets that are larger, have a larger sweet spot (which can dramatically improve both quality and consistency when hitting the ball) and generate more power and more spin. Simply put, modern technology can make the bad player better and the good player better still.

But there is one significant difference between the use of performance enhancing technology in business and in sport. In sport, performance enhancing technologies are strictly controlled. The rapid improvement in racquet technology meant that the International Tennis Federation had to create new rules regarding the design of racquets to limit their impact on the game. In complete contrast, there is no international governing body that limits the use of performance enhancing digital technology in business. In most cases they are actively encouraged because business is a competitive environment just like sport.

There are occasions when performance enhancing technologies are banned or discouraged, but worryingly this is normally done by the organisations that would benefit from the outcome. Not only do organisations often fail to provide the best possible technology to their users, some also have strict policies on people bringing their own technology into the workplace. This is often done on the fallacy of protecting the companies from malicious intent but most often only serves to limit the effectiveness of our top performers[x].

Keeping with the analogy for just a little longer, in sport there is also often an explicit strategy by your opposition to target your weaknesses. If you have a poor backhand in tennis then your opponent will try to hit to your backhand whenever possible. In cricket, if you have a preference to hit to one side of the field the opposing team will try to bowl to make that difficult for you.

In business, we have weaknesses as well. It turns out that one of our biggest weaknesses when it comes to thinking work is our ability to remember things accurately. Perhaps as a result during my school years, and even through my university days, memory was considered a key cognitive skill. It was even tested through a cruel, and in my mind pointless, process called closed book exams. In the real world it is not the remembering of information that is considered an important skill, it is the ability to find it, and ultimately apply it, which is considered valuable. Today, most of us use either digital tools, such as laptops and smartphones, or even non-digital tools, such as paper and pens, to assist us with our memory.

Without these types of tools, our ability to remember things is flawed. Not only is there a tendency to forget things that actually happened, we also create false memories for things that never

---

[x] I call this a fallacy because examples such as WikiLeaks and Edward Snowden's disclosure of classified documents show that even the CIA and NSA are incapable of protecting themselves from malicious intent. The focus on risk reduction rather than performance goes back to the misalignment of the IT Department's objective of security with the organisation's objective of performance.

occurred. Research conducted by Utrecht University on returning Dutch war veterans highlighted this phenomenon. Two months after their return from Afghanistan, the veterans were given subtle misinformation about a fictional missile attack on their base. Seven months later at a subsequent debrief, 26% of participants had come to believe that the fictional event was true[25].

A 2012 article in *New Scientist* titled 'Memory: Remembrance of things to come'[26] suggests that the reason our memory is such a mess is that our brains were never intended as memory engines. They are more suited to the creative process of imagining scenarios for the future than recalling things from the past. To generate these scenarios our brains create a collage from past events, effectively cutting them up and stitching them together again. Once this has been done it is incredibly difficult to put back together the original memory again.

In the world of modern business when the information that we rely on is often digitised or digitisable, it makes sense to use our digital tools to enhance the performance of our memory. Additionally, more recent research out of the University of California, Santa Cruz, suggests that the ability to save information digitally results in a type of cognitive unloading. It frees our brain's capacity to be used for other things, either other memories or potentially for other cognitive endeavours such as the capacity to think, solve problems, and generate new and creative ideas. We try to remember information in the belief that it will have a future value to us, but as researchers Benjamin Storm and Sean Stone put it, "By treating computers as extensions of memory, people may be protecting themselves from the costs of forgetting while taking advantage of the benefits."[27]

Although most of us have had access to personal computers for decades it is only in the mobility era (since about 2005) that opportunities of having enhanced memory have come to fruition. The mobility era has provided us with the opportunity to capture and save all our important memories digitally. We carry in our pockets the technology to be able to record conversations, take notes, mark up documents and take photos. Prior to this we may have had a selection of digital memories but only alongside a much larger collection of analogue ones (stored in our notebooks and on paper) and cognitive ones (stored in our brains). Our ability to recall any one of these memories when we needed it, or to even remember that we had it in the first place, was reliant on our flawed memories.

Looking for information in an analogue world is a time consuming and human intensive activity. It requires searching and flipping back through pages of a notebook, going through filing cabinets, searching bookshelves and leafing through books. The quantity and variety of information that we need to manage in business means we waste huge amounts of time looking for information. In its 2008 research article 'Enterprise Findability Without the Complexity', Google suggests that knowledge workers spend more than a third of their time searching for information (either successfully or unsuccessfully)[28]. Even allowing for a certain level of bias in the results, this is still a stunning amount of time to spend on an activity that is ultimately of zero value. Yes, looking for information is of zero value.

The zero value activity of looking is related to the high value outcome of finding but there is little relationship between the amount of time you spend looking and the quality of the outcome. If someone is completely disorganised and takes three hours to find the report you are looking for then the report is still no more valuable than if they had found it in five minutes (given that information has a tendency to go out of date, you could argue that by getting it three hours later it is worth marginally less).

When we operated in an analogue world, we may have believed that there was a strong relationship between the act of looking for information and the act of finding it – that the time spent looking was inevitable. What's more the premise of looking for something was preceded by the idea that you could remember what you were looking for in the first place. But today, time spent looking for something is an avoidable expense. Digital technology allows us to process complex information easily, as well as find things quickly, including things we might otherwise have forgotten.

But the true cost is the lost opportunity of doing something else – something value-generating. We can unload our brains so that we can spend more time on higher value cognitive tasks and the things we love to do. In the competitive world of business, digital technology provides a performance enhancing edge[xi].

[xi] This list is not exhaustive. For more information, I highly recommend (again) Clive Thompson's book *Smarter Than You Think*.

DIGITISE WHAT YOU CAN AND FOCUS ON EVERYTHING ELSE

As we approach the end of this section, I hope for two things, or maybe three. The first is that you see that the digital transformation currently happening in business is about you. Sure, it is also about macro trends and organisational strategies but the pointy end of the stick is going to be felt by people.

Secondly, I hope you now see that in this inevitable shift there is both a challenge and an opportunity. The challenge comes if you choose not to change; in this case you will gradually become worth less and perhaps one day be 'rationalised', with elements of your job reassigned to a robot or computer program. But I believe this outcome is neither desirable nor inevitable. There are actions we can take that will not only protect us but potentially grow our future relevance and our future value. The opportunity lies in choosing to act, choosing to proactively engage with and tap into the amplifying power of digital technology. This is the path that will maintain your relevance, and give you the knowledge and skills to increase your value by trading on your uniquely human skills.

There will be those who can see that the cost of inaction is far greater than the cost of action, but still won't act – just like quitting smoking or saving for your retirement, for some the benefit will be too distant and the change too hard. And there will be others who point to Moore's law and the ongoing improvements in digital technology as a reason to delay action. They will argue that by waiting until tomorrow, the technology will be cheaper and better than it is today.

But both of these approaches are misguided. Like quitting smoking and saving for your retirement, the longer you delay action, the smaller the benefit and the further behind everyone else you are. And so the third thing I hope for is that you have developed a desire to take action, to embrace your digital future and leverage the advantage it can give you. The fundamental question we need to address is not what we can protect but rather, in the words of Salim Ismail, "what else can be information-enabled?"[29]

The next section is for people who have a desire to take action. In it I will share with you all that I know about helping people transition to digital technology. With this guidance, I hope you will find the strategies you need to digitise what you can and free yourself to focus on everything else, everything that is more valuable, more enjoyable and more human in the workplace.

REALISTIC

DON'T DISCOUNT SUCCESS

So you're still with me? I hope that means I have laid out a compelling argument as to the inevitability, importance and opportunity of digital technology (or you may have skipped the first couple of sections, in which case you will just have to take my word for it).

Much of the reasoning I have provided is extremely logical; it involves chunking your work into individual tasks, allocating the right ones to technology, managing information correctly, being more effective and multiplying returns. But for some reason, even with an overwhelming abundance of logic many people have been slow to adopt technology or still actively avoid it. For years I been making this same argument as to why we need to be using more digital technology and using it better, and how that will allow us to make more informed business decisions which will impact positively on the world we live in, yet even close friends and family have failed to act.

I cannot tell you the level of frustration that this has resulted in for me. How is it that even people who trust you implicitly, who know what you are saying makes sense and who know they really should do something about technology, fail to engage in meaningful learning and change?

It has only been quite recently that I have realised that the decision to act is not necessarily based on logic. Regardless of the size of the opportunity the decision to act is as much about our perceived chance of success as it is about the rewards. Perhaps our perceived chance of success is even more important than the reward itself as we will discount or downplay the importance of the rewards if a successful outcome seems unlikely.

This is in some ways similar to going to the casino, where the games with a lower chance of success need to offer better payouts to entice you to play. On a roulette wheel if you bet on a single number you have a one in 37 chance of your number coming up. Because the odds of success are so low the payout is 35 times your bet as a reward if you win.

If we believe that success is either hard or unlikely then we discount both the relevance and the potential rewards. As a result, we will spend our valuable time elsewhere and create reasons to justify our inaction. So how can we shift the odds in our favour? How can we make success seem more realistic and ultimately tap into the opportunities we have discussed earlier?

Just like the casino, not all bets are the same. Some games are pure chance, like roulette, and others such as blackjack and poker are based on both chance and skill. With games of skill there are strategies that you can use to dramatically improve the chance of success. In blackjack if you follow a strategy where you keep taking cards until you bust then you can be guaranteed of failure 100% of the time. If you follow a strategy that reflects the rules the dealer has to follow, you reduce the failure to about 8% of the time, and with a more complex strategy you can reduce that to about 0.5% of the time (depending on the table's specific rules).

Success using technology is not a game of pure chance, it is a game of skill. If we want to learn to use digital technology effectively, we need a strategy for success. So what does a successful strategy look like and how can it be implemented? From my experience there are three key ingredients to a successful strategy, and bringing them together is like cooking. Not baking, where you need to get your proportions just right and follow the instructions exactly, but cooking where you can experiment with changing proportions and even adding in a few extra ingredients, playing with it until it tastes just right.

The first and most important ingredient for successful change is individual mindset. You need to believe that change is both desirable and achievable before anything else. The second ingredient is environment. You need conditions that actively encourage and support learning and growth. And the third ingredient is process. You need an approach to learning that minimises effort and maximises success.

I believe that the key recipe to successful digital learning programs is having the right mindset, the right environment and the right approach together. I would go so far as to suggest that the failure of digital learning programs can be directly attributable to one of these three things not being adequately addressed. Where is my evidence, you may ask? I believe the best possible evidence for this is our kids.

In workshops and training programs I often hear participants comment how good kids are with technology. But they aren't born with some sort of super power – they didn't pop out of the womb already clutching a remote control. Yet where we often seem incapable of even mundane technology tasks such as setting the time on the microwave or programming the VCR, we immediately expect that the average teenager can do it with ease.

So what is it about the mindset, conditions and approach to learning children experience that allows them to excel with technology? In the next few chapters you are going to find out. Over the rest of this section we are going to look at these three ingredients in more detail and then formulate a strategy to maximise our chances of success.

THE GHOSTS OF TECHNOLOGY PAST

In her book *Mindset: The New Psychology of Success*, Stanford University professor Dr Carol Dweck distinguishes between two different mindsets that can have a dramatic impact on our ability to learn and develop. A fixed mindset is one in which you believe your intelligence, personality and other traits are largely fixed and will not change over time, while a growth mindset is one in which you see that your abilities can continually be developed and improved.

Although we may have a general disposition towards one or the other we can also have different mindsets based on different activities or areas of expertise. So in one part of your life you may continue to grow and expand your knowledge and in others you might be quite resistant to change.

Although Dr Dweck's research has shown that having a growth mindset is a driver of long term success, success may not necessarily result in a growth mindset. In fact success in one part of our lives might contribute to a fixed mindset in another part. If you perceive that your level of success or achievement is an important part of your self-image then you will act in ways that will protect it. As Dweck puts it:

> *From the point of view of people with the fixed mindset, effort is only for people with deficiencies. And when people already know they're deficient, they have nothing*

> *to lose by trying. But if your claim to fame is not having any deficiencies – if you're a genius, a talent, or a natural – then you have a lot to lose.*

What this means is that as we get older, our achievements in one area of our lives might undermine our decision to embrace growth and learning in another, especially if we perceive a high chance of failure. And when it comes to digital technology we may perceive that our chance of failure is higher than it really is. Older generations grew up in an era where technology was a lot less reliable and a lot less usable. As a result many of us are holding onto ghosts of technology present that make us unwilling to actively explore the opportunities that new technology offers.

The ghosts of technology present are the mental scars that were inflicted during the early stages of digital technology. They include the memories of losing an important document or assignment only hours before it was due to be submitted, painfully slow dial up internet speeds, or the 'blue screen of death' on Windows (or the 'spinning pinwheel of death' for Mac users) that signified an imminent computer reset and the loss of hours of work.

These ghosts, these failures, encourage a fixed mindset, because they reinforce a belief that we just aren't good with technology and this encourages us to remain disengaged. Many of us have grown wary and fearful of digital technology and the fear and apprehension that sits just below the surface means we only use technology when we absolutely have to.

Although digital technology has come a long way since the dark days of the 1980s and 90s, we still find the odd bit of evidence to support our previous fears and in doing so reinforce a fixed mindset. We download a poorly designed app and are reassured that technology is still unintuitive; we drop our tablet, smash the screen and blame the build quality; we forget to charge our phones and see this as evidence that battery life is still unacceptable; and yes, once in a while our device will still crash for inexplicable reasons. Each one of these experiences reinforces our fixed mindset and reassures us that our decision to abstain is a sound one.

But anyone under the age of 20 largely missed the PC era altogether, and their first computer was likely to have been a smartphone or tablet. As a result they missed the 20 most frustrating years of computing[xii] and they are largely unaware of the ghosts that continue to haunt us. Additionally, they don't have a reputation to protect. They aren't yet considered experts or professionals and they don't have staff, colleagues or customers that they need to constantly impress.

There is one final factor at play when it comes to the different mindset our children hold towards digital technology, and it is that they don't necessarily know worse. We often think that children act up and make mistakes because they don't know better – they haven't learnt the 'correct' behaviours, the 'right' way to do things. But when it comes to technology, this is not the case, it is that they don't know worse. They don't have old, established ways of doing things that hold them back. Consequently, they are free to play with and embrace digital technology and tend to find ways to use it to enhance their lives.

Over years or even decades in the workforce, along with a decade or more of formal schooling, we develop a system of information management that works. This might be as simple as a pen and a moleskin to write things in or it might involve some intricate

---

[xii] It is not that computing wasn't frustrating before the 1980s, it was just that far fewer people had to deal with it. Therefore, if you take the level of frustration and multiply it by the user base, the early 1990s was probably the most frustrating era of computing to date.

combination of highlighters, lever arch folders, file dividers, and those little coloured tags (I can attest to having tried both these systems during my years of formal education). Once we have a system that works we don't want to change it, even if what 'works' only works in comparison to having no system at all. We stop looking for better and we settle for the status quo.

If dial up internet and brick size mobile phones are the ghosts of technology present, it is lever arch folders, archive boxes, overflowing filing cabinets and 'notebooks that we left in our other bag' that are the ghosts of technology past. These are the ghosts of the analogue age that still haunt us but that we couldn't opt out of like we could with digital. So it seems that our kids have avoided not only the ghosts of technology present, but also the ghosts of technology past.

So are we without hope? Is the mental scarring so bad, our sense of self so misplaced and our lives so jaded that we can never recover? Or is there hope that we can develop a growth mindset like our children? According to Dr Dweck the single most important factor in moving from a fixed to a growth mindset is knowing that the two mindsets exist. By acknowledging that you might have a fixed mindset and being able to identify the fixed mindset voices in your head, you open yourself up to choosing a different path.

An important aspect of developing a new mindset is recalibrating our view of what success means. With a fixed mindset, success is often about what we have already achieved and the status that we have. With a growth mindset success is about what we have managed to overcome. By associating success with the process of learning and development rather than the outcome, you are more likely to engage in new ideas that you might have otherwise avoided. We are only at the beginning of our digital journey and a growth mindset will encourage us to continue to learn along the way.

Finally, we need to recognise that many of our beliefs are stories that we tell ourselves to make sense of the world or to justify our actions. Although our ghost stories about digital technology were likely relevant and accurate when we created them, there is a good chance they are now past their use by date.

An exercise that I often run in workshops involves participants writing down their negative beliefs about digital technology (on paper if they have to) and clarifying what actions that leads them to take. I then ask participants to consider if the opposite belief was true, how would their actions be different? People are often surprised to find that many of their beliefs are already questionable, and with the rate of technology change there is a good chance that the other ones will become questionable soon.

A mindset is like a muscle that needs stretching. Our capacity for change and adaptation depends on how often we've done it before. The risk is that we start to lose that muscle if we continually accept the status quo and resist growing, if we don't identify or act on the opportunities that present themselves even if the risk is sometimes failure.

There is nothing that will stop the rate of technology growth continuing to accelerate. With a fixed mindset digital technology will go past us like we are standing still, but with a growth mindset we have a chance of at least keeping up, if not moving into the lead.

A LEARNING ENVIRONMENT

Back in high school I studied French for about four years. I was never that interested or engaged in the classes, and I found it all a little frustrating. Quite predictably my French never quite made it to the conversational stage, but I didn't mind because as a 15 year old I couldn't see where I was going to use it again in the future.

It turns out that about five years later I was living and working in the UK and had a number of French friends. I went to visit them in France a couple of times and both my short-sightedness and my limited French became all too apparent. I could scrape by when purchasing a train ticket or ordering a baguette but after four years of formal learning I was still unable to hold a meaningful conversation in French.

So why did I find learning French so hard? At the time I used to point to the irregularities of feminine and masculine nouns as being the culprit but looking back it is clear there were other things at play. Firstly, I had a fixed mindset when it came to French. I was fairly sure that I wouldn't be good at languages and subsequently managed to prove it to myself.

Secondly, the learning environment didn't necessarily support the activity that it was designed for. Learning French at high school consisted of going to a language laboratory once a week for a face to face lesson and then going home to do self-directed learning (which meant doing exercises in a workbook with the help of a French-English dictionary).

There was no one in my family who spoke French, so through this approach to learning I only got feedback on my progress once a week. In addition I was practising things without the opportunity for real world application and I found it difficult to be motivated. The outcome of all this is that I probably spent much of the four years relearning the things that I had forgotten from the weeks, months or years before.

Of course there are other more effective ways of learning a language. One approach you can take is to go and live in a foreign country. This 'throw yourself in at the deep end' (or immersion) approach creates a significantly different learning environment. Firstly, the importance of the skill you are learning is reinforced on a daily basis, and the learning serves a clear function. The inability to communicate effectively will at best mean social exclusion and at worst a monotonous diet of baguettes. Secondly, being surrounded by native speakers means that you have easy access to peer support when you need help or make mistakes.

I think that this is a fairly solid analogy for the differences between how we and how our kids learn about digital technology. Our kids have learnt in an environment where the use of digital technology is common and understanding is seen as valuable. Children interact with their digital tools on a daily (or hourly) basis meaning that their learning is constantly reinforced and it sticks. And they do their learning surrounded by their peers, people who will support them when they make mistakes or need help.

In contrast, older generations like to learn irregularly and in isolation. Often, we don't have access to the right expertise, and the skills we are learning aren't necessarily seen as valuable by either ourselves or our peers. For a long time people who openly use technology have had to make excuses for using their digital tools and developing their expertise.

I cannot tell you the number of times I hear executives question the use of digital technology by others as a form of self-reassurance about their own use of paper. This response is largely driven by the fear that the geeks might be onto something and they risk being left behind (it's okay, 'geek' is considered a compliment these days). But rather than see that as an opportunity to grow and learn it is easier to reject and downplay what others are doing.

In reality the technology avoiders should be the ones making excuses. Generally they are less efficient and less effective, place a disproportionate burden on shared resources (such as IT services and administration) and reflect worse on the organisation.

I had a client recently that worked for a mortgage broking firm and one of the mortgage brokers insisted on using a paper diary. As a result the support staff were always calling her and chasing her up to check her availability whereas they could check the availability of the other brokers in real time. Because the other staff couldn't check her availability they also couldn't confirm appointments in real time with clients. This was not only a productivity issue but also one of reputation.

Using a homing pigeon these days would be considered absurd, even sending a fax would raise an eyebrow or two. Isn't it time that we put all our paper in the same waste-basket (so to speak) or at a bare minimum, give respect to those that are looking to push the boundaries of work?

So how can we improve our learning environment? Many, especially those working in larger organisations, might see this as being outside their control. Some organisations have a strong bias towards protecting the status quo and as such have a culture, and often even written policies, that discourage experimentation and change. But even in these environments we can still effect change in our immediate environment. We could form a peer technology group and arrange a regular time to catch up (at least weekly). This time could be used to share the things that are working well for you and get help and support. Perhaps once a month someone could research and demonstrate a new app that they think the others might benefit from. And talk to kids – most likely they will be encouraged that you want to talk to them about digital technology rather than trying to kick them off it.

We all have the ability to impact our learning environment on a personal level, even if it means doing so outside of our work hours. Of course if you have the authority or ability to influence the learning on a broader level then I would encourage you to act on it. We need to be explicit about our support of digital technology in organisations and to embed the changes that will create positive learning environments not just for ourselves but also for others.

In one of the most successful mobile technology programs I have worked on, the organisation formed a mobile projects team. The team was multi-disciplinary with strong expertise in IT and training and it answered directly to the CIO, bypassing the head of IT. One of the first things they did was start capturing real life stories of people using technology in the organisation and sharing these with everyone they met. Through this explicitly supported demonstration of peer-level experiences, they created a new norm for technology use and influenced the culture of the organisation.

LEARNING THAT STICKS

The third ingredient of successful learning programs is our approach, and once again it seems there is much we could learn from our kids.

Over the last 20 years of digital technology, the most common approach for technology learning, especially within organisations, has been osmosis. The way this approach works is that when we are sitting near people who we think are competent, we look over their shoulder and copy whatever they are doing. Sometimes we might be so bold as to ask a direct question, but often this is avoided lest we appear unknowledgeable.

In hindsight, it is not surprising that this approach doesn't appear to have been very effective and if you need any evidence you need look no further than email. If you were to ask a room full of executives whether they considered themselves to be effective with email you would probably get about one in ten raise their hands. If you were then to ask the same group whether they had ever had any formal training on email you would also be lucky to get a one in ten response.

The problem is not just email. People feel overwhelmed and ineffective with so many of their digital tools, and usually they have had little if any formal training. The modern workplace often just expects that everyone is good with technology, that people intuitively know how to use digital tools, even though our personal experiences might tell us the opposite is true. This belief has led to a situation where we not only systematically undertrain our staff, but also indirectly promote that the poor use of technology at work is okay.

The first problem with this approach is that it is not really an 'approach', and it means most organisations generally don't have a structure around digital learning and development. You might argue that this is no different from our children, that they also don't generally have a formal approach to learning or access to technology training opportunities, but there is a significant difference. Whether it be formal or not most of their life is one big learning opportunity. At school or at play, life is rarely serious enough that a child can't afford to experiment or try something new. But at work in the adult world, the opposite is true. Things are almost always too serious and we can rarely afford the time to play and learn. In most roles, if we want to learn we have to carve out a specific amount of time and pay a trainer to come in to justify the time we took out of our day job and have 'evidence' to demonstrate the value of undertaking that learning activity in advance.

And therein lies the second problem with our approach. Even if we do have access to formal learning, the method most organisations follow to purchase training does not necessarily promote the right outcomes. Much like IT, training is often seen through the lens of compliance and cost, and is designed to meet arbitrary measures to do with cost per participant rather than long term value and outcomes. This is because the people tasked with organising the training are rarely the people who experience it and also rarely the people who suffer if it is bad or benefit if it is good. This is especially true in larger organisations where there is often a separate learning and development department with a budget for training and development activities, but it can also be true of smaller organisations where menial tasks such as sourcing training providers are given to the lowliest of administrative assistants.

When training is purchased, and therefore provided, with an eye on cost rather than on value you generally get training that is abbreviated, functional and standardised. This type of training is shortened to be cheap, is made generic by focusing on how technology works (rather than what it does and how that can help you) and is standardised to make it replicable (so it can be delivered by the lowest common denominator of trainer).

Take a moment to consider an alternative approach. Instead of abbreviated, functional and standardised training with an eye on cost we could have expanded, actionable and customised training focused on value and learning outcomes. This type of training would be more spread out over time to embed learning and would promote long term change. It would focus on how technology could be applied in a way that is valuable to participants and it would be customised to make it meaningful and relevant to the learner's specific context.

Expand your training

Perhaps the bigger challenge to overcome is the investment of time that is required. People are increasingly time poor and organisations are now pushing more and more for shorter training. Ironically they will often pay more if you can reduce your day long training program to three hours or your three hour program to a breakfast session. The premise behind this is our people's time is valuable and while they are being trained they are not doing value generating work.

To me this type of approach echoes the use of diet pills and those Power Balance holographic bracelets as an alternative to doing regular exercise to improve your physical fitness[xiii]. It is an approach borne from a pursuit of quick fixes rather than willingness to do the work that is required to create long term meaningful change. This is not to say that you can't do better training that is shorter, but a focus on the length of time rather than outcomes is rather pointless.

Once again we need to move out of the cost paradigm and focus on value. The question shouldn't be how long will this take, but rather how many hours will I get back for each hour I invest. This return on time invested is not just a great way of measuring the impact of the learning experience but also a wonderful motivator if you are an overworked executive or business owner. When I work with my own coaching clients and ask them what they will do with the time they could save each week, the answer is never 'more work'.

Expanded training is not just about total time but also about spreading the time out. In most cases even three hours of intensive learning would contain more information than we could possibly take in. If we were to instead spread those three hours out over a number of weeks, and in between provide opportunities to apply the new skills you have learnt, then you are not only more likely to remember, you are also going to get an immediate return on your time invested.

[xiii] It claimed to use holograms to "resonate with and respond to the natural energy field of the body", resulting in an increase in sporting ability. Although endorsed and used by elite sports people such as basketball player Shaquille O'Neil and racing car driver Rubens Barrichello the manufacturers were eventually forced to withdraw any claims relating to their impact on physical performance due to an unsurprising lack of evidence.

Make it actionable

Perhaps my single biggest gripe with more traditional IT training is a focus on function rather than action. To make the training generic and applicable to a whole classroom full of people (thus reducing the cost) the trainer will focus on how the technology works, providing a tour of its menus, rather than focussing on what people can use it for.

I had the displeasure of doing an Advanced Excel course a few years ago and one of the big features of the course was pivot tables. The trainer diligently stepped through the process of creating pivot tables and even gave us a set of dummy data to work on. The problem was that this wasn't my data and it wasn't a problem that I was dealing with right then. I wasn't learning by doing something relevant to me or to fulfil a function, and as a result my current recollection of pivot tables is about zero.

Instead of starting with the technology and then trying to find a problem to practise it with, find the problem and then start looking for the suitable technological solution. This will make the learning experience immediately meaningful, actionable and valuable. This approach will also hopefully avoid some of the worst effects of Shiny Object Syndrome, where we waste time playing with new apps and other shiny objects that distract us from our work rather than amplify our impact.

Customise your learning experience

Perhaps the greatest success of the online learning world is the Khan Academy. Started by Salman Khan and supported by the Bill and Melinda Gates foundation, the Khan Academy provides free access to around 6,500 videos on subjects ranging from mathematics to computer programming to American history.

Although these videos are generic and functional rather than actionable, Khan Academy does a great job of letting you personalise your learning experience. After your initial assessment you can skip content that you have already shown proficiency in and you can move through the content using different pathways so you can focus on what is most interesting and relevant for you. But perhaps the biggest advantage of Khan Academy is the ability to watch videos and undertake example problems for as long as you need before you move on.

In his 2011 TED talk "Let's use video to reinvent education", Salam points out that one of the biggest problems with the way we educate our kids is that it is the small things they miss and don't understand that undermine their confidence and ability. Because the teaching continues to move forward whether each student grasped the last concept or not, learners end up with 'Swiss cheese gaps' in their learning[30]. By customising learning experiences and giving individuals the ability to slow down, speed up and change course at their own pace, students can learn more and learn the things that matter most to them.

As I don't work in learning and development, I can't comment on whether this type of approach is a reasonable or radical proposal, but it does closely align to the way our children learn to use technology, and they seem to be doing a far better job of it than most of us. They don't go to a half day intensive learning program with no follow up; instead they have short but regular learning interactions over a longer period of time. This gives them the opportunity to learn, practise and apply new ideas and explore the fringes of their knowledge before the next learning cycle starts.

With much of our formal training we have the opportunity to learn and practise new skills, but without any chance for real world application until after the training has been concluded. Any questions that might arise after the formal learning component often go unanswered because there is no one to ask them of, and subsequently learning stops.

The limitations of formal learning are well documented. Research conducted in the 1980s in California showed that formal learning amongst teachers led to new skills being applied in the classroom only 10% of the time. When the opportunity for practice and feedback was added it still resulted in less than a 20% rate of implementation. Interestingly, when peer coaching was added and provided the opportunity for learners to be supported during application, new skills were implemented 90% of the time[31].

The type of approach that I am advocating requires the development of peer experts that can be a part of the learning journey with you. As discussed in the previous chapter most people generally lack a large pool of digitally adept colleagues to choose from and this may be seen as somewhat problematic. It also implies a significantly larger time commitment to learning than many organisations (and perhaps even individuals) are comfortable with.

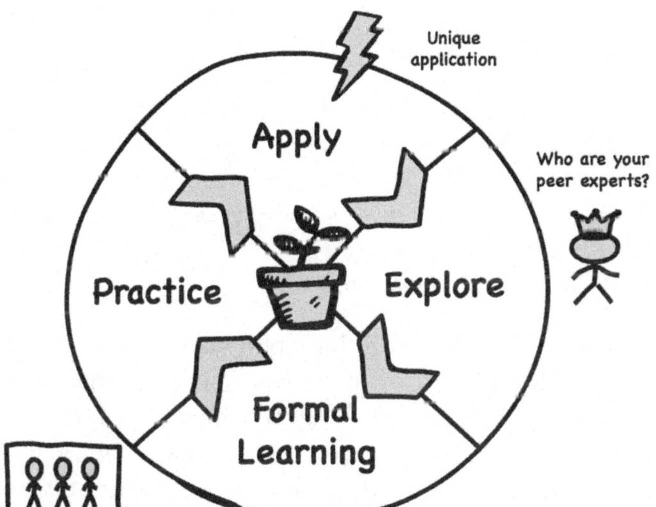

These should not be seen as limiting factors but rather as hurdles that need to be overcome. The time commitment needs to be seen in the light of investment rather than cost and given the right time and resources there is no reason why peer experts cannot be cultivated in organisations. For small business owners and sole practitioners there is also an emerging cohort of digital coaches that will work with you one to one to develop your knowledge and skills. The ever forward thinking Seth Godin pointed out the need for digital coaching way back in 2008 in his blog post titled, 'A shortage of digital coaches'.

> *Here are three things that are true:*
>
> *1. Digital technology, especially computers and cell phones, can dramatically increase productivity.*
>
> *2. More and more users of digital technology are small firms or individuals.*
>
> *3. The vast majority of users of digital technology are totally lame in getting the most out of the investment of their time and money.*
>
> *There is still much to be learnt from the approach that our children take to learning technology…This is a growth area. The tools are finally in place to market yourself, and even better, the productivity increases you can produce for $500 or $1,000 are not just valuable, they're actually priceless.*[32]

Unfortunately it has taken the best part of five years for the world to catch up with Seth but over the last few years there has been an emerging field of digital coaching. To be clear, I also work as a digital coach and mentor and it would therefore be easy for you to dismiss my statements in this section as an act of self-promotion. But the reality is that the ad hoc approach we have pursued over the last 20 years is clearly not working, and in a world of accelerating technology use this approach is no longer acceptable. We have got by with this approach to date not because it was effective, but because in a world of analogue business it wasn't so important – if you were good at business you could get by with not being good with technology. But now more so than ever before we need to be good at both to be successful.

START WITH SIMPLE SOLUTIONS

I can feel your excitement. Your mindset is set to growth mode and you have surrounded yourself with peer experts to go on this learning journey with you. But there is still one more problem: with so many options you don't know where to start. On the one hand the pervasiveness of digital technology provides so much opportunity but on the other we can often be paralysed by both the level of complexity and the choices.

This problem is like a friend giving you a hard drive full of music. The digitisation of music has now made it possible for hundreds or even thousands of songs to be shared on a whim. The problem with this apart from the obvious copyright infringement issues is that it is incredibly difficult to separate the good music from the bad. If we were to add all this music to a playlist we could still be working through listening to the tracks a day, a week or even a month later.

Trying to find the artists or the songs you really like within this stream of music is near impossible, firstly because the music was never filtered with your preferences in mind, and secondly, you don't get a chance for individual artists or tracks to sink in before you move to the next track, and the music just washes over you. In days past a friend might have given you a mix tape or a burnt CD. You weren't given as much music but each track mattered. You were given music that was tailored to your preferences and you had time to listen to it. It was easier to find music that mattered.

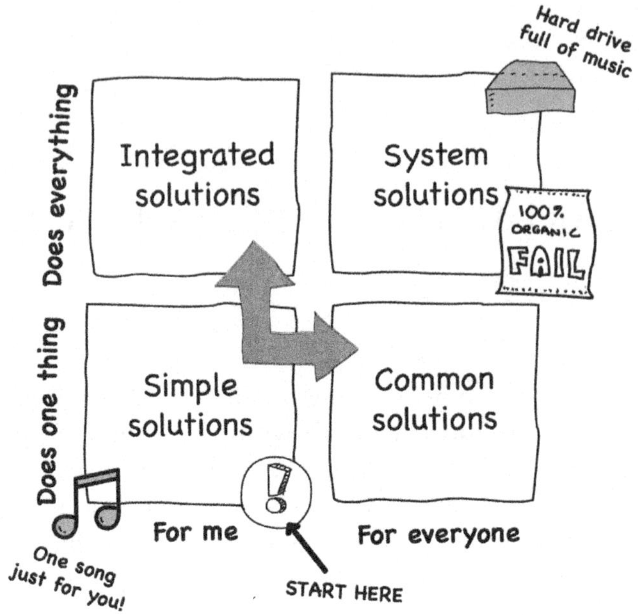

So the answer to the question of where to start with digital technology is: start small and simple. We want to focus on finding simple solutions to meaningful problems and then break down the learning into small, manageable chunks.

In my mind simple solutions have two key characteristics. Firstly, they are focused on the individual rather than the organisation – they are customised for the problems of one person and work in a way that is suited to that individual. Secondly, they are task focused, aiming to achieve one specific outcome, without necessarily needing to integrate with other technology.

For a long time implementing simple solutions wasn't possible. Especially in larger organisations technology was largely standardised with no consideration for individual needs or preferences. You were provided a core set of applications, normally Microsoft Office, and the installation of non-standard software was largely prohibited. To make this solution suitable for the wide range of different roles in the average organisation, each of these programs was very generic in nature.

The popular saying known as 'Maslow's hammer' states, "if all you have is a hammer, everything looks like a nail", and in the past this very much applied to the way our technology worked. It was like having a tool kit that consisted of a hammer, saw, flathead screwdriver and adjustable spanner, except the toolkit was Microsoft Office, and in the same way everything looks like a nail when all you have is a hammer, everything used to look like an Excel problem when all you had was Excel. Need to create an invoice? That's an Excel problem. Want to add up some numbers? That's an Excel problem. Want to track your customers? Clearly an Excel problem. Want to create an advanced business model? Another Excel problem (actually, that really could be an Excel problem).

But with the arrival of the iPhone as well as a new breed of online services, this era has thankfully come to an end. The development of smartphone and web apps means we now have a multitude of tools to choose from and we can get one specifically tailored to the task at hand. It is like moving from the tool kit that you keep under the kitchen sink to what you might find in a mechanic's workshop. A mechanic is likely to have a full set of imperial and metric ring spanners, a socket wrench set and a torque wrench just for tightening up nuts and bolts, and likewise a modern knowledge worker can now easily have a range of specialised tools.

This 'appification' of work and the ability to choose a specific app for a specific task has a number of advantages. Because modern apps and services are more specific in their use, they are often much simpler than the more generic tools of the past which had to accommodate everyone. The options and functions contained in the app are restricted to ones that are specifically required for a particular task.

Additionally, if an app is generic with many functions you need to have the expertise to apply it, whereas with specific apps you can tap into the expertise of the developers. For example, if you use Excel as a way of tracking customer interactions it won't guide you as to which fields or bits of information you should be tracking and it won't suggest when or even if you should reach out to them regularly. In comparison a modern Customer Relationship Management (CRM) program will provide text fields that guide you on which information needs to be collected, it will provide a funnel or some form of pipeline that indicates how you should segment your current leads and some will even auto-assign follow up tasks to make sure you keep in contact.

Each app has the opportunity to provide a single, simple solution to a task that we want to perform. We should look to have one app per task and one task per app. By focusing on one task and one app at a time we can dramatically reduce the overwhelming burden of learning new technology.

Apart from engagement and ease of learning there are a number of other advantages to this approach. We only need to invest a small amount of time before we start to generate a return; and at least some of the knowledge and confidence we build learning one app is transferrable to subsequent apps. If an individual app doesn't meet your needs you can just get another one – you don't need to replace the whole system.

Although simple solutions can make learning easier and they are an excellent way to start your digital learning journey, the eventual shift to system solutions should ultimately be seen as desirable. There will still be occasions when we need more complex system solutions. Tasks will still need to be rolled up into activities and these activities will still need to be aligned with larger business objectives. This will require our simple solutions to be integrated together and also that certain solutions are used by everyone. Integrating solutions with each other can improve productivity by seamlessly moving data and information to where it is required and common solutions can allow for the sharing of knowledge and information across an organisation. What doesn't work is when organisations start with a system approach and end up with a highly capable solution that no one is capable of using.

So if we can't necessarily have each user choosing to use a different document format or having them implement their own preferred CRM system, where do we start? There are a number of tasks that we currently undertake that are neither integrated nor common: the tasks we carry out in analogue form, or more specifically, the tasks we do on paper.

Each sheet of paper works in isolation. Apart from being stapled together with other sheets (which can easily be overcome) it is in no way integrated with any other system, and although we may all use the same brand of notebook we all use our own methods and approaches.

So the best place to start with digital technology is with your analogue activities. Think about the activities you do on paper (taking notes, creating task lists, reading and marking up documents). It is no coincidence that there are 100s or even 1,000s of apps available to help you do each of these tasks faster, cheaper and better than you can do them on paper. Tasks you do in analogue are the best opportunity for you to begin your journey to digital.

MAKE CHANGE A HABIT

In the words of management guru and author of *Leading Change*, John Kotter, we need to see that "transformation is a process, not an event". Learning how to use technology is a journey that we will go on, not a destination we will someday reach. Although we've seen many advances, digital technology is still in its infancy and we don't truly know what is in store next. A successful approach to change should prepare us for the present and also continue to support our growth as technology continues to evolve.

Fundamental to a successful approach are the ingredients that we have discussed through this section: a growth mindset; a supportive learning environment and an approach focused on long term engagement and action. But even with all this in place we are still relying on good intentions to drive your learning – that after you put down this book you will have the motivation, willpower and discipline to follow through and implement. The reality is that good intentions will probably not get you where you want to go.

Rather than good intentions we need to be able to rely on something else, something that requires altogether less effort, and that is the opposite of intent: habit. A habit is an acquired behaviour pattern regularly followed until it has become almost involuntary[33]. Habits take a whole lot less energy, willpower and motivation than other actions. For us to be successful on our journey towards technology mastery we need to make change a habit rather than an isolated incident.

According to Charles Duhigg, author of *The Power of Habit: Why We Do What We do, and how to Change*, our brains are inherently lazy and habits allow it to conserve energy. The good news is that the human brain will try and turn almost any repeated behaviour into a habit if it can, and to do so it requires three key components: a trigger or cue that indicates the start of the process; the routine or 'chunk' of behaviour we want to undertake; and a reward at the end.

Our ability to create habits around technology are already well proven, it is just that most of them are less than constructive. Take for example the ring of a mobile phone. When a phone rings (the cue), most people feel compelled to drop what they are doing and answer it. The act of answering the phone is the chunk of behaviour that is largely done without thinking; we pull the phone from our pocket or bag with single handed dexterity, swipe or press a button and put the phone to our ear in one swift movement. The reward at the end of this is a sense of social inclusion that comes from talking to someone who wanted to talk to us.

For some people the habit is so well formed and the behaviour so automatic that they fail to make any consideration for other environmental factors that might make such behaviour unacceptable, such as being in a meeting, driving a car or even having sex[xiv].

[xiv] This appears to be a common question to ask on surveys in the UK. Depending on which survey you believe somewhere between 10% and 25% of people have taken a call or read a text message during sex. For more information you can check out these two sources (accessed 8 March 2015): http://www.centralsomersetgazette.co.uk/Shocking-statistics-reveals-South-Westerners-use/story-20849387-detail/story.html,

Thankfully the same mechanisms that create bad behaviours can be used to create good ones. All we need to do is find the right cue and reward and associate it with a positive behaviour. Rather than coming up with new cues and rewards we can use existing ones. For example, we could replace the behaviour of reading the paper when we have our morning coffee with watching 'how to' videos on YouTube. Sitting down in the cafe is the cue and the caffeine hit becomes the reward.

Another option is to enlist the support of others to help motivate you. Peter Cook, author of *The New Rules of Management: How to Revolutionise Productivity, Innovation and Engagement by Implementing Projects that Matter*, explains how statistics show that you are more likely to be obese if people you know are obese, primarily because it consumes too much willpower to act outside your social norms. As Cook points out, if you want to get fit, hang out with fit people. Similarly, if you want to learn about digital technology, hang out with other people who want to learn – or people who have already learned. In Cook's words, "find your tribe".

> *Why not create a weekly 'digital explorers' lunch with a bunch of likeminded colleagues who want to test out new apps? The cue for learning is lunch and the reward is socialisation and shared learning. In addition, when your own motivation might be running low you can tap into the shared motivation of your tribe.*

---

http://www.yourtango.com/201165808/shocking-stat-25-percent-people-answer-phone-during-sex

Let's finish on some good news: once habits form they stick. Eventually your lazy brain will take over and it will be easier to do the good things, just as it was once easier to do the bad. Given that there is no end in sight for our digital journey, doing the hard yards and setting up the right behaviours now means that in the future you will be able to adapt to ongoing changes, and apply your willpower and motivation to the other things that matter in your life.

WHAT HAPPENS NEXT

OUR DIGITAL FUTURE – PART I

As I have mentioned ad nauseam through this book, we are only at the beginning of our digital journey. Whether it be as individuals, organisations or at the scale of society, digital technology will continue to reshape our working lives at a much faster rate than in the past. Trying to get our heads around this is sometimes difficult but the following excerpt from an interview with technology commentator Kevin Kelly provides some insight into the magnitude of change we can expect.

> *If we were sent back with a time machine, even 20 years, and reported to people what we have right now and described what we were going to get in this device in our pocket—we'd have this free encyclopaedia, and we'd have street maps to most of the cities of the world, and we'd have box scores in real time and stock quotes and weather reports, PDFs for every manual in the world—we'd make this very, very, very long list of things that we would say we would have and we get on this device in our pocket, and then we would tell them that most of this content was free. You would simply be declared insane. They would say there is no economic model to make this. What is the economics of this? It doesn't make any sense, and it seems far-fetched and nearly impossible. But the next twenty years are going to make this last twenty years just pale. We're just at the beginning of the beginning of all these kind of changes. There's a sense that all the big*

> *things have happened, but relatively speaking, nothing big has happened yet.*[34]

We have no way of predicting with any certainty what the next 20 years of digital technology will bring. The things that seem impossible today may become reality in only a few years. But we don't need to understand the technology of tomorrow to get a glimpse of the future; we only need to look at the technology of today. As author William Gibson pointed out, "The future is already here, it is just not evenly distributed yet." To understand our digital future all we need to do is extrapolate from technologies that are emerging elsewhere.

There are many books out there already that explore the digital trends of the future (please see the Further Reading section for some of my recommendations). The focus of this book has instead been on how we engage practically with digital technology in the present. However, it is worth taking just a small amount of time to consider what is on our immediate horizon, if for no other reason than to reinforce the need for constant learning and change.

In my opinion the three most important and exciting shifts we are seeing for technology users in business are contextual computing, natural user interfaces and hyper-connectivity. These shifts do not exist in isolation – their roots can be traced back to more fundamental trends in the technology landscape, such as the growth of networks, increasing computing power and the ongoing reduction in computing costs. But these fundamental trends do little to inform us of the impact that end users will experience. There are of course other shifts that could equally warrant further explanation, but let's start by looking at each of these three shifts in more detail.

First up, consider contextual computing (or context aware computing). An agreed definition of contextual computing is hard to come by, but you could do worse than the definition offered by Gartner:

> *Context-aware computing is a style of computing in which situational and environmental information about people, places and things is used to anticipate immediate needs and proactively offer enriched, situation-aware and usable content, functions and experiences.*

This means that our devices and apps are using information such as where we are (based on GPS location), what we are doing and who we are with (based on calendar and contact information) to prioritise information and activities that are most relevant and demote others which are not required.

A good example of this is the Google Now service, which will scan your email for flight confirmations and then on the appropriate day, and based on your current location, inform you of when you need to leave to go to the airport, the best route to take, traffic conditions and up to date departure information for your flight. In business this type of approach is starting to emerge with CRM systems that integrate with Google Apps for Work (effectively Gmail for business users) and provide customer information based on who the email is from.

The value of contextual computing is derived from two core inputs, the amount of information that we have digitised and the ability of our devices to correctly understand our current situation. As digital technology use becomes increasingly prevalent, the quality and quantity of content will continue to grow. In addition, as more of our physical and behavioural landscapes are mapped, the ability for our digital tools to uncover patterns of contexts will also improve.

There has also been a long term trend towards more seamless interaction with technology, which brings us to the second shift: natural user interfaces. Starting with the computer mouse and graphical user interfaces in the 1980s, there has been an ongoing quest to reduce the friction between ourselves and technology and over the last decade there has been a lot of progress. The use of gestures such as touch, pinch and swipe in mobile technology have dramatically reduced the friction of using technology. With the advent of technology such as Microsoft's Kinetic and the Leap Motion controller, interaction with technology has now moved off the screen and can be performed in three dimensions.

Over the last few years we have also seen a dramatic improvement in the ability of voice recognition software to understand both content (what is said) and context (what is meant), and early stage facial and emotional recognition with eye tracking is now built into several mobile devices such as the Samsung Galaxy. This provides an additional context to enable our technology to appropriately tailor content and functionality when we require it.

If that isn't exciting enough, we are now seeing the early stages of direct brain-computer interfaces. Companies such as Australia's EMOTIV are producing consumer headsets that read the electroencephalography (EEG) signals from the brain. By mapping thoughts to actions the headset can be used to interact with video games and television sets, and even fly drones and remote control helicopters, using just thoughts and facial movements.

With these developments, digital technology usage is set to become much less cumbersome, but we will still need to engage with and learn some of these new interfaces as they emerge. In its report "Being Human: Human-Computer Interaction in the Year 2020", Microsoft points out that, "This trend is the end of interface stability." Although the ability to touch type and use a mouse will still be important skills in the immediate future, they will continue to be marginalised and replaced[35].

The third trend that I see as being of particular relevance to business users is hyper-connectivity. Hyper-connectivity refers to the use of multiple means to enable human to human and computer to computer communication. To some extent we are already dealing with it, as most of us are already contactable via phone, email and SMS, and many of us could add a Twitter, Facebook or LinkedIn account to that list.

I understand that for many people already overwhelmed by the connectivity of today, the increased connectivity of tomorrow probably doesn't sound appealing. But as we have discussed throughout this book, every technology presents both a challenge and an opportunity. Generally the challenge arises if we are not prepared for the change and take a reactive rather than proactive approach. We have a choice in how we react to these trends, but we don't necessarily get the option to stop them.

Hyper-connectivity between computers has the opportunity to automate many of the information flows that are currently manual. Whether it be the flow of your credit card data to your expense reporting program or the automatic assigning of email and phone call information to your CRM system there is the potential for hyper-connectivity to simultaneously improve the quality and reduce the physical burden of information activities.

The opportunity of hyper-connectivity is to allow information to flow freely from where it is created to where it is needed. Our input will be to decide where we want it to flow to and which channel we want to access it through.

Within these three shifts – contextual computing, natural user interfaces and hyper-connectivity – we are seeing a fundamental change in our relationship with digital technology. For much of human existence our relationship with technology has been at arm's length, literally. Whether it be an axe that augmented our strength, the use of calculators to improve our maths, pens and paper to augment our memories, our tools have all been operated by our hands and fingers. This was fine when the technology was the 'dumb' part of the process but the growth in computing power and a shift towards more thinking and less doing work has changed the dynamic. Both our brains and our technology now have the capacity to work faster than our hands, and this means it is our physical self that is now the weakest link in the chain.

As Clive Thompson points out in his excellent article "The Joy of Typing"[36], transcription fluency, or the ability and speed with which we can transcribe our thoughts, is a key factor in both our creativity and the quality of our work. In the article he points to research into the effects of typing speed on student performance. The studies found that students who typed their assignments rather than writing them using a pen "increased their ability to produce text that was creative and original, more technically accurate, more logically sequenced and better organised and showed greater pragmatic awareness and sensitivity to audience"[37].

Typing is a great example of trying to overcome the weak link between our thoughts and our ability to capture them, but even skilled typists can no longer keep up with the speed of technology if the interaction is at arm's length. Starting with mobile technology we are now developing an even more intimate relationship with our technology in a bid to overcome this weak link. This intimacy is being developed in both a physical sense and a cognitive one; our devices are in closer proximity to us and we are relying on them more and more for key activities such as communication, memory and cognition.

A good way to describe this developing relationship is the same way we define a cyborg: "a person whose physiological functioning is aided by or dependent upon a mechanical or electronic device"[38]. For many cyborgism has been the realm of science-fiction – Arnie as the Terminator or the Six Million Dollar Man, but cyborgism is much more real and perhaps much less threatening than its fictional representation.

Cyborgism doesn't necessarily require technology be built into our bodies, as pointed out by Ben Hammersley, author of *Now for Then: How to Face the Digital Future without Fear.* He argues that when computing power doubles every two years, no one is going to want last year's plug embedded in their head. So apart from quality of life situations such as pacemakers, cochlear implants and artificial limbs, this somewhat crude form of cyborgism is unlikely to get a lot of traction.

So rather than building the technology into our bodies, we will start carrying and wearing our technology more and more. I would argue that this form of cyborgism is not only possible, but in the new world of work it is also highly desirable. Although the only jobs that are completely digital are those done by computers, there is a greater proportion of *every* job, a growing number of tasks, which are now conducted virtually. This means the more seamlessly we can integrate the activities we undertake in the physical world with the information available in the digital world, the more effective we will be at work.

As pointed out earlier, unlike other competitive activities such as elite sport, there are no rules against performance enhancing technologies in business. The search for an edge in business will create a demand for an increasingly seamless integration between people and technology. Each of these will require a period of adoption, but like email, electronic calendars and smartphones they will eventually become a part of the business landscape.

I have used the term cyborg to be intentionally confrontational, but if our digital future is filled with technology that is both useful and easy to use, then I believe that cyborgism is not just an outcome that we will see as acceptable, it will be something to be desired. It will be a pathway to a future of greater humanity by digitising and automating the tasks and activities that don't play to our strengths.

Without a doubt, this is a rose-tinted view and it is not a future that we can take for granted. There is another path that we are also following, one in which technology is competing for our attention and often winning, one in which technology is used against us and has the opportunity to suck the humanity out of us. There is another digital future we need to consider next.

OUR DIGITAL FUTURE – PART II

In 1998, Linda Stone, formerly of Apple and Microsoft, coined the term 'continuous partial attention' to describe the ongoing fight we have to protect our attention in an age of technology.

> *To pay continuous partial attention is to pay partial attention — CONTINUOUSLY. It is motivated by a desire to be a LIVE node on the network. Another way of saying this is that we want to connect and be connected. We want to effectively scan for opportunity and optimize for the best opportunities, activities, and contacts, in any given moment. To be busy, to be connected, is to be alive, to be recognized, and to matter.*
>
> *We pay continuous partial attention in an effort NOT TO MISS ANYTHING. It is an always-on, anywhere, anytime, any place behaviour that involves an artificial sense of constant crisis. We are always in high alert when we pay continuous partial attention.*[39]

Every technology user has dealt with continuous partial attention. It is the type of problem that makes the idea of greater technology integration and cyborgism sound just a little bit scary. Although there is an opportunity for technology to create a positive future, there is an equally possible future where opportunities are squandered and we are sucked into a meaningless digital abyss.

For anyone who has wasted hours browsing the web or playing meaningless games, or lived with distracted teenagers (or adults) incessantly checking social media at the dinner table, you are right to be concerned about the possibility of a digital future that takes away what is good and meaningful and gives us little in return.

Just as there are developers and inventors out there working to create a positive digital future there are others with much more sinister motives. There are people out there competing for our eyeballs, plotting to take our attention away from whatever else we could or should be doing. They will use every trick in the book and exploit our psychological weaknesses in a bid to suck up our time and ultimately relieve us of our money. Want an example? Look no further than the innocuous sounding world of social gaming.

Social gaming sounds fairly harmless, and using a broad definition it has been shown to have positive impacts like promoting social interaction, cooperation and creative thinking. This applies to social gaming in its broadest sense – games that you play with your friends, either in a room together or over the internet.

There is another, more sinister breed of social games that I believe are much more concerning and have the potential to be much more damaging. Games such as Zynga's FarmVille, FarmVille 2, CityVille, and CastleVille, or King's Candy Crush, Candy Crush Saga, Candy Crush Soda Saga and Pet Rescue Saga have more wrong with them than just unoriginal titles. These types of games are being intentionally designed to consume our attention.

Academic and game developer Ian Bogost provides a great critique of these types of social games and the potential damage they can cause. To prove both the pointlessness of these games and their powerful effect he created the somewhat satirical Facebook game called 'Cow Clicker' where the sole objective of the game was to log on every six hours and click on a cow.

This may not seem like much of game, but that's the point. Like many social games there is no point – there is no meaningful objective, just a bunch of routine actions that encourage you to both enlist your friends and ultimately part with your money. Our engagement with these games is often on the pretext of 'zoning out' and 'downtime' but the outcomes are much more insidious than a lost hour or two.

In Bogost's blog post, "Cow Clicker, The Making of Obsession"[40], he points out that in addition to the time that is destroyed there are a number of other (and most often intentional) negatives. Firstly, the nature of these games impacts on real life social relationships. By allowing you to promote the game to your social network in exchange for in-game currency, your social network is reduced to a resource to be exploited rather than relationships to be respected. These games also actively encourage compulsion (using similar techniques employed in slot machines), with the primary objective of getting people to spend money. Thirdly, many social games require minimal skill and as a result provide little, if any, meaningful reward for progressing through the game.

For these types of games the term 'social' is misleading. There is little interaction with others except posting your progress in the game to social networks (which is little more than advertising), and many of these games are played in isolation and as a distraction from what is happening in the real world. Just as in casinos, these games have an intentional strategy to disconnect you from the real world for as long as possible so as to more easily separate you from your money.

To give you an idea of the size of the potential problem, King Games, maker of Candy Crush Saga, generated revenue of nearly US$2.3 billion from social games in 2014[41]. This equates to about half the combined revenue of all 23 casinos on the Las Vegas strip in 2013[42].

Aside from the financial waste on what is ultimately a meaningless and empty pursuit, it is the time wasted that is perhaps most concerning. Research on casual game players (people who play these types of social and mobile games) shows that the vast majority spend between one and four hours playing games each week[43]. Even if we were to take the low end of this range, one hour per week, and multiply it by the 93 million active users of Candy Crush Saga, we would find that this one game alone accounts for over 4.8 billion hours of wasted attention each year.

We now live in what Michael H Goldhaber calls the attention economy. All economies are based on the trade of scarce goods and in the modern developed world one of the few things that is still scarce is our attention[44]. Access to food, clothing, shelter, entertainment and education have all become more abundant, yet we all still have only 24 hours of attention that we can give each day. Businesses are now seeing that your attention is a gateway to your wallet and whether it be social games, Facebook, or television there is a lot of competition for this scarce resource.

Building on the idea of the attention economy, John Hagel, Co-Chairman of the Deloitte Centre for Edge Innovation, points out the need for us to consider our 'return on attention' (or ROA), or how much benefit we derive from each moment of our attention that we give[45].

This is why these types of social games represent the worst of technology – they consume millions of hours of attention and there is little, if any, return on your attention investment, and for many individuals there is a clear financial cost. This is technology at its easiest and most useless. Compared to this I would prefer my mid 1990s PC. It might have been harder to use but at least it was useful. And the fact that it was hard meant that we were generally unlikely to use it unless we had something meaningful to do. Sure we had games back then but you didn't see people wasting 4.8 billion hours on Frogger or Sim City.

Why have I spent so much time talking about games? Wasn't this supposed to be a book about technology in business? Well, I believe that the use of digital technology in our personal lives and our professional lives is intrinsically linked. How we use technology at home has a massive impact on how we use it at work, and vice versa.

These alternate digital futures are both equally real and both already here. They are two sides of the same coin and will be part of the landscape that we need to navigate. How we navigate it, and the destination that we reach, will depend a lot on the actions of people like you. Disengagement is not an option in the long term; sooner rather than later you will have to decide whether you will be a part of the problem or part of the solution.

THE HUMANITY OF TECHNOLOGY

The longer I have been working with people and technology, the more I have realised that the challenges that need to be faced are generally not technology ones. Over the medium to long term we have shown an ongoing ability to exceed many of our wildest dreams when it comes to technology development. The challenges that we face, and will continue to face, are human ones.

I think there are two broad challenges that we particularly need to consider. Firstly, what is the big picture purpose of using technology, and secondly, how do we want technology to work. I describe the first as the Technology of Humanity and the second as the Humanity of Technology.

The Technology of Humanity can be summed up in one simple quote often attributed to Uncle Ben: "With great power must also come great responsibility." Benjamin 'Ben' Parker was Peter Parker's (a.k.a. Spider-Man) uncle and surrogate father. His words of wisdom were instrumental in ensuring that Peter applied his powers for good rather than selfish endeavours.

Modern digital technology gives us what is akin to mental superpowers, and it is time that we stopped using them selfishly or wasting them altogether. We need to use our superpowers in a way that amplifies the human condition and contributes to a better future for not just ourselves but also our families, society and future generations.

Each one of us already carries with us the computing power required to put a man on the moon. Just a few decades ago scientists would have paid millions of dollars to access the computing power that we can purchase for a few hundred dollars. Yet rather than put it to good use, most of the time it sits idle, or is used as a distraction from more important things.

We can easily contribute more to the human condition, for example by putting our phones onto charge. Projects such as BOINC operated by the Space Sciences Laboratory at Berkley University allow users to contribute their unused computing power, including the power of your idle (charging) mobile phone, to the crunch data for global problems such as finding a cure for HIV or improving our understanding of climate change. Currently the BOINC network ranks as having about the same computing power as the fifth fastest supercomputer in the world.

As the performance of technology continues to grow and our superpowers grow with it, we will need to become increasingly conscious about how they are used. If we want the best out of technology then that is what we will need to create. Over the longer term, improvements to both the quality (and longevity) of human existence is ours to create or destroy.

The second challenge, or the Humanity of Technology, is a path that we have thankfully already started down. If I was going to sum this up with a Spider-Man quote it would be, "Right now, I'd trade the whole Spider-Man bit for a rocking chair and a subscription to *Reader's Digest*."[46] For a long time, doing useful work with digital technology has been hard and we need to start making it easier.

For instance, think about how many badly designed PowerPoint presentations you have seen. If your experience is like mine, the answer would be most of them. Although some of this responsibility clearly falls on the author, it is also true that PowerPoint has an incredibly complex menu structure with hundreds or perhaps even thousands of options to tweak fonts, transitions, backgrounds, animations, borders, transparency and colours, to name but a few. This level of complexity is overwhelming for most users and results in people either not doing things they should because it's too hard or doing things they shouldn't because they don't know better.

Compare that to Haiku Deck, an image intensive PowerPoint alternative that actively guides users to create high impact presentations. Not only is it much simpler to use than PowerPoint, it also encourages greater use of visuals, fewer words and consistent formatting, all factors that have been shown to contribute to more effective presentations.

I think there has been a general belief that meaningful work needs to be hard, that somehow the effort is intrinsically linked to the value of the outcomes. Thankfully this belief is now being questioned and principles of design and games are now being increasingly built into the tools of meaningful work.

This is an area of rapid improvement but there is still much work to do. The emerging field of gamification is one example of how we can build engagement and usability by applying game principles to the way that we work. Dr Jason Fox, author of *The Game Changer: How to Use the Science of Motivation with the Power of Game Design to Shift Behaviour, Shape Culture and Make Clever Happen* points out that the underlying mechanics of games – goals, rules and feedback – are all present both in our work and more broadly in our lives. By making meaningful work goal-driven, challenge-intense and feedback-rich we can encourage better outcomes. The fact that digital technology provides a never before seen platform for infusing work with these elements (much like in computer game development) means we have an opportunity to do this like never before.

Whether we have a future in which technology works to further humanity or reduce it is in our hands. Ultimately we get to choose the technology that we want to use but to make good choices we need to develop digital understanding, or a digital intelligence. In the absence of informed users we will continue to be force fed a high sugar technology diet by developers who have no interest in our long term wellbeing. Just like you need to know a bit about nutrition to have a good diet, you need to know a bit about digital to determine the good from the bad and the useful from the useless, and to apply technology in a way that benefits not only us but also the world around us.

WE NEED DIGITAL INTELLIGENCE

The premise of this book was how to avoid extinction in a world of digital business. Given the risks that technology brings, it is reasonable to question whether a digital future is desirable at all. Unfortunately this is a moot point; whether we have a digital future is not a choice, it is the inevitable outcome of human progress and is impossible to stop. The choice is not between a digital future and an analogue one; the only choice is whether we want to play an active role in creating the type of digital future we want.

How do we simultaneously move towards a future of greater digital engagement and at the same time avoid going down the rabbit hole? Especially when the game is being played against multi-billion dollar app and game developers (and other purveyors of shiny objects) who are collecting mountains of data and constantly testing new ways to capture your attention?

Often we believe that the answer to stopping using something badly is to take it away, but perhaps the solution to combating the worst of technology in our personal lives is to use the best of technology at work.

In 2013 the American Society of Anaesthesiologists conducted research on the impact of digital technology on medical students. The students were given iPads to use as study aids. The study found there was a high uptake of using the digital tools, with about

88% of those students given the devices using them as part of their studies. A surprising outcome of the study was that as students increased their use of devices for educational purposes, their level of personal use dropped. Over a six month period personal use of their iPads dropped by about 50%[47].

This suggests that the more we use something well, and for its intended purpose, the less novelty it has and the less likely it is to negatively impact our personal lives. The best strategy to protect against the worst of digital technology is to spend time using the best of it at work.

The inverse of this is also true: overusing the worst of technology at home negatively impacts our use of technology at work. The way that we use technology in our personal lives is often extremely passive. We don't feel a need to justify our actions, make good technology choices or limit distractions. It's our downtime and we can spend it how we want.

But after hours lost playing mindless games or browsing Facebook, we are likely to perceive technology as both time wasting and useless. We are also likely to underestimate our own ability to use it well, because it is impossible to use it well when we are not doing anything useful with it.

Whether it be in our professional or personal lives, we need to develop the knowledge and skills to better understand how digital works. Our ability to acquire and apply new knowledge and skill is what we call intelligence and we are going to need a new form of intelligence, a digital intelligence, to be able to survive our digital future. This is something that needs to extend beyond the workplace and should be developed from an early age. It is just as relevant to our family and our children as it is to us and our role in business.

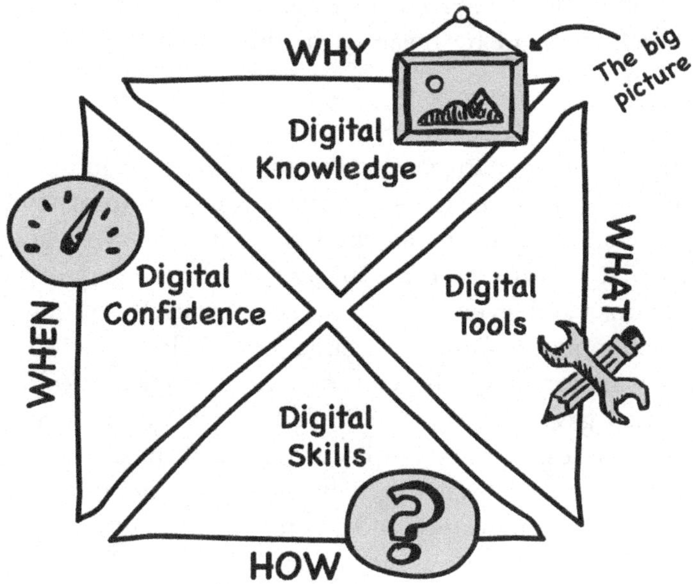

The digital intelligence that I talk of is an innately human ability and is not to be confused with artificial intelligence or machine learning (where the term digital intelligence has also been applied). Digital intelligence is our ability to acquire, and ultimately apply, new understanding and skills relating to digital technology. It is more than the ability to use digital tools – it is the know why, know what, know how and know when of using digital technology to improve effectiveness and outcomes. Digital intelligence is about our relationship with technology, just as emotional intelligence is about our relationships with others.

Now, I appreciate that the concept of intelligence is an ambiguous one (and I also accept that there are far more qualified people than I to discuss it), but it is clear that our concept of intelligence has changed over time. In the early 19th century our understanding of intelligence was dominated by concepts of logic, pattern recognition, numeracy and literacy. Whether by coincidence or as a direct result of business need, a strong understanding in these areas was in high demand by business in the middle of the Industrial Revolution.

And although logic, pattern recognition, numeracy and literacy still form the basis of IQ tests, our most common form of intelligence measure, we now accept that there are other areas of knowledge and skill that we also require. During the 20th century our view of organisations, and even society at large, started shifting from a mechanical one to a more organic one. Whereas 19th century business was dominated by factories and the making of stuff, the 20th century gave way to more thinking work and the managing of people. As a result the workplace moved from a low to a high human contact environment. Again it is probably no coincidence that during this time we saw the emergence of competing forms of intelligence measure such as emotional intelligence (EQ).

As we enter an age of work and life that is increasingly conducted using digital tools, it is appropriate that we once again recalibrate our concept of intelligence. It is here that the concept of digital intelligence (DQ) fits, not at the exclusion of IQ or EQ but as a logical addition to them.

Digital intelligence has four key elements. The first is to understand why we would want to use technology, its strengths and the opportunities to apply it to our advantage. The second is knowing our options, what technology is out there and the ability to choose the right tool for the job. The third is understanding how it works and having the ability to apply our digital tools in an effective way. Finally, we need to develop the judgement to know when technology should be used, when it is going to benefit what we are doing and when it is going to subtract.

Of these it is perhaps the fourth, judgement, which is the most important. Ability in itself is not enough, as it only gives you the how; judgement is something developed through a combination of ability and experience. It is for this reason that many people who are extremely competent with technology, including our kids, often fail to exercise judgement as to when it is appropriate to use it. Just because you can, doesn't mean you should.

Ability combined with experience, and especially diverse experience, provides a much better basis for sound decision making than ability alone. The Oxford Dictionary definition of judgement is "the ability to make a considered decision and come to sensible conclusions"[48]. Considered decisions require the ability to know and understand alternatives and this in turn requires diverse experience. It is probably not surprising that the average age of a judge in Australia's High Court is close to 65 years old, considering they are making important decisions about how to apply the law – an area where we want people with both the right knowledge and the right experience.

It is also worth pointing out the difference between experience and diverse experience. Most of us have experience with email but that hasn't necessarily resulted in good judgement. As our understanding and ability with email grew we went through stages from 'I won't' to 'I can't' to 'I should' before finally getting to 'I can' and 'I will'. Most of us have been stuck at 'I will' for the last 15 years without ever developing the judgement of 'I might'. 'I might' requires the knowledge of alternatives and a high level of ability to allow us to critically assess the other communication options. Thankfully, after 20 years of experience the conversations about 'I might' and whether we should use email are finally coming to the surface.

Developing digital intelligence is about short-cutting this process. It's about developing the knowledge and skills that will allow us to understand new technologies as they emerge, to identify the opportunities in them and manage the risks. Digital intelligence is transferrable, and it is as relevant in our personal lives as it is in our professional ones, just as IQ and EQ also cross those boundaries.

Most of us have already started developing a basic DQ but in a contest with game and app developers where the big prize is our attention, relevance and perhaps even our sanity, our current ad-hoc approach to developing DQ isn't going to cut it. Purposeful and structured digital learning is imperative for anyone growing up or growing old in the 21st century. As Kevin Kelly pointed out, we have no idea what the technology 20 years from now will look like. But we can be sure that there will be a lot more of it and our ability to use it well will be more important than ever.

BE A LEADER WITH TECHNOLOGY

I wrote this book for you. I wrote it to help you understand what the digital revolution meant at an individual scale, how all the talk of driverless cars, robots in the workplace and the endless parade of technological advancements impact on people in the workplace. I wrote it in part to provide clarity about what is going on, about the opportunities and risks, and in part to provide motivation and direction for what you could do next.

But this isn't just about you. The digital revolution is not isolated to your job or your industry and it is not even isolated to the world of work. Its first tremors are already being felt by millions if not billions of others. It will impact your colleagues, it will impact your family and friends and it will impact society at both a local and global scale.

Given this, we need more than just individual action. We need to mobilise whole businesses and organisations. We need to share what we know and provide an environment for others to learn in. We need to create a movement towards the better use of technology and that is going to require leadership. We are going to need leaders with technology.

A leader with technology uses technology to maximise their leadership traits and at the same time sets an example of better technology use. The leader may benefit individually but this is secondary to the impact they can have by influencing others.

If we are to consider some of the key leadership traits there is ample opportunity to use digital tools to amplify our leadership. Although there seems to be as many leadership models as there are leadership books, there are some common domains of activity that leaders undertake. These include maintaining a clear vision and direction, facilitating communication and collaboration, enabling people and overseeing process. I appreciate that this is an incredibly simplistic view of leadership and that under each of these domains there are various activities and responsibilities. Given that this book is primarily about technology, and not about leadership, I hope you don't mind if I keep the discussion fairly high level.

In each of these domains there is the opportunity for digital technology to amplify our leadership performance. Our ability to maintain a clear vision and direction is improved because digital technology amplifies our awareness. It moves information further and faster than ever before possible. Our understanding of global events is not limited to the stories that appear in our local paper but instead by real time feeds direct from the source.

Digital technology allows communication and collaboration to occur point to point, bypassing the traditional hierarchy, and it can happen in real time regardless of geographical distances. Digital technology is also making communication more open, making the right resources easier to find and knowledge easier to access.

As discussed throughout this book, digital technology can be a great enabler of people, but potentially a disabler as well, however research suggests that a more open approach to digital technology generally has a positive impact on people. A joint study by Google and Deloitte Access Economics showed that a flexible IT policy which allowed people to bring their own devices to work and use technology remotely meant that workers were one-third less likely to leave their job[49].

Finally, the integration of digital technology into so many of our organisational processes means that we are able to get real time feedback in a way never before possible. Digital sensors and real time data analytics mean that the monitoring and management of processes is no longer limited by geography or time.

I appreciate that the above is well short on detail, but expanding on each one of these would provide enough material for another book. The objective of my brief summary is to highlight the opportunity that effective use of technology provides to amplify our leadership activities.

As a word of caution, I believe that the image of technology has suffered from years of hype that it could only ever fail to live up to. I am not proposing that digital can 'amplify your leadership or your money back!' Rather, used well there is the opportunity to improve your performance and also set a visible example of digital in action. Through application we will learn and many of these learnings will come through mistakes, but even with these risks I believe that our digital future is still both bright and inevitable.

Like most of us, I have experienced the negative impacts of technology and confess to spending many hours in a state of continuous partial attention. I have dealt with the frustrations of technology that didn't do what it said on the box and I still check my phone far more often than I should. Yet after all of this, I have no doubt that digital technology has amplified my communication, my memory, my level of organisation and my situational awareness. Technology is just a tool, and it is the decisions we make (or fail to make) that define whether technology works effectively for us or not.

For a long time technology advocates have remained silent or made excuses for their use of digital tools, but it is now time for them to raise their voices, not just for themselves but for the good of their organisations and also ultimately for society at large. Perhaps the defining feature of digital leadership is a deep seated and vocal conviction in a positive digital future and our role in creating it.

Given the fairly dismal use of digital technology and a general ambivalence in most organisations, this may seem like an insurmountable challenge to overcome, but if we want a future defined by the best of technology and not the worst then we have little choice. The good news is that according to recent research it only takes a small group of committed people to change the majority opinion. A study by Rensselaer Polytechnic Institute in the United States showed that as little as 10% of a population coming to hold an unwavering belief, and being committed and vocal advocates, can change the opinion of the majority. The research paper cites historical precedents for this including the suffragette movement in the early 20th century, and the rise of the American civil rights movement[50]. In every society-level change, there is a tipping point that can lead to dramatic changes in beliefs and culture, but for this tipping point to be reached it is important for leaders to vocally and unwaveringly show support for their cause.

We sit at the cusp of one of the most significant changes in human history. Calling it a digital revolution is not just hype or propaganda, it is the most accurate description we have for what is happening in the workplace. We need to choose the role that we will play in the revolution. Will we be an advocate for old technology and old ways of doing things? Will we risk being on the wrong side of disruption? Will we be able to adapt to a new digital order, to take advantage of what it has to offer us and manage the challenges? Or will we step up and go on a journey to shape the digital future, not just by ourselves, but taking others with us?

THE LAST LITTLE BIT

Thank you for making it to here. I appreciate that your attention is valuable and there are many books that you could be reading, and your commitment of time to this one is an honour that I hope I have done justice to.

Most importantly, I hope that I have shown you how digital technology is relevant to how you work as well as how it can enhance your performance and amplify what it means to be human in a world that is increasingly digital. Finally, I hope that I have managed to show you a way forward – how approaching the way you learn with the right mindset, in the right environment and with the right approach can make learning not only easier but more enjoyable.

Digital technology is providing opportunities to work in ways that were previously unbelievable and it will continue to revolutionise the way we work in the future. This is not without its risks but the better we use our digital tools the better equipped we will be to deal with them.

It is now time for you to decide what you are going to do next. Very soon you will put down this book and be faced with a decision. A decision as to whether this book was infotainment or inspiration. A decision as to whether you want to keep doing what you've always done and slowly get less than you've always got, or whether you want to take action to prepare you for the digital workplace of the 21st century.

None of us are complete novices anymore but many of us are still amateurs and being a professional in the 21st century means making a decision to invest the time, the effort, and perhaps even the money to learn the digital tools of your trade. This is the decision you need to make.

For you to stay the course, your decision needs to be rooted in a deep seated desire for a better future, both for yourself and for those around you. For me, this desire is also tinged with a dissatisfaction in the status quo, a belief that our present relationship with technology is unhealthy. This desire and dissatisfaction are the catalysts for action – they are the pieces of grit that will keep irritating you until you do something about it.

But desire alone will not get you there. You need to believe that a future where we have a positive and rewarding relationship with digital technology is possible. If we do not believe that a better future is possible, and within our ability to influence, we will fail to take meaningful action.

I do not pretend to say that change will necessarily be easy. In fact, I can guarantee that at times it won't be. What I will guarantee is that the long term rewards will make the short term challenges worthwhile.

And now is the time to make your decision. Our old ways of work are going extinct and you need to work out whether you are willing to go down with them. If not, you need to make a commitment to action, to learn how to use digital technology and to embrace the opportunities of our inevitable digital future.

FURTHER READING

Ashton, David, Brown, Phillip & Lauder, Hugh. *The Global Auction: The Broken Promises of Education, Jobs, and Incomes.* New York: Oxford University Press, 2012.

Beinhocker, Eric. *The Origins of Wealth: Evolution, Complexity, and the Radical Remaking of Economics.* Boston: Harvard Business Review Press, 2007.

Brynjolfsson, Erik & McAfee, Andrew. *The Second Machine Age: Work, Progress and Prosperity in a Time of Brilliant Technologies.* New York: W. W. Norton & Co. Inc., 2014.

Cohen, Jared & Schmidt, Eric. *The New Digital Age: Transforming Nations, Businesses, and Our Lives.* New York: Knopf Doubleday Publishing Group, 2013.

Ford, Martin. *The Lights in the Tunnel: Automation, Accelerating Technology and the Economy of the Future.* USA: Acculant Publishing, 2009.

Ismail, Salim. *Exponential Organizations: Why New Organizations are Ten Times Better, Faster and Cheaper than Yours (And What to Do about It).* New York: Diversion Books, 2014.

Kahneman, Daniel. *Thinking Fast and Slow.* USA: Macmillan, 2011.

Kurzweil, Ray. *How to Create a Mind: The Secret of Human Thought Revealed.* New York: Penguin Group, 2012.

Thompson, Clive. *Smarter Than You Think – How Technology Is Changing Our Lives for the Better.* New York: Penguin Group, 2014.

NOTES

---

[1] S McPherron, 'Evidence for stone-tool-assisted consumption of animal tissues before 3.39 million years ago at Dikika, Ethiopia', *Nature*, vol. 466, August 2010, pp. 857-860.

[2] C J Lepre, 'An earlier origin of the Acheulian', *Nature*, vol. 477, 2011, pp. 82-85.

[3] M Hilbert, 'How much information is there in the "information society"?', *Significance*, vol. 9(4), August 2012, pp. 8-12.

[4] Hilbert, *Significance.*

[5] H A Simon, 'Designing Organizations for an Information-Rich World', in Martin Greenberger (Ed.), *Computers, Communication, and the Public Interest*, Baltimore, The Johns Hopkins Press, 1971, pp. 37-72.

[6] H Hazlitt, *Economics in One Lesson*, Harper & Brothers Publishers, New York and London, 1946.

[7] C B Frey & M A Osborne, 'The future of employment: How susceptible are jobs to computerisation?', academic publication, Oxford University, Oxford, 2013.

[8] R Kurzweil, 'The Law of Accelerating Returns', Kurzweil AI Network, 2001, retrieved 22 February 2015, <http://www.kurzweilai.net/the-law-of-accelerating-returns>.

[9] 'The office of the future', *BusinessWeek*, vol. 2387, 30 June 1975, pp. 48-70.

[10] '2045: The year man becomes immortal', *Time*, vol. 177(7), February 2011.

[11] J Rosen, 'The knowledge, London's legendary taxi-driver test, puts up a fight in the age of GPS', *NY Times Magazine*, November 2014, retrieved 22 February 2015, <http://tmagazine.blogs.nytimes.com/2014/11/10/london-taxi-test-knowledge/>.

[12] J Federman & M J Rosenthal, 'Waze sale signals new growth for Israeli high tech', Yahoo! News, June 2013, retrieved 22 February 2015, < http://news.yahoo.com/waze-sale-signals-growth-israeli-high-tech-174533585.html>.

[13] R Florida, 'Building America's third great job machine', Creative Class Group, 2011, retrieved 22 February 2015, <http://www.creativeclass.com/_v3/creative_class/2011/05/05/building-america%e2%80%99s-third-great-job-machine/>.

[14] E Rosen, 'Every worker is a knowledge worker', *Bloomberg Businessweek*, 11 January 2011, retrieved 22 February 2015, <http://www.businessweek.com/stories/2011-01-11/every-worker-is-a-knowledge-workerbusinessweek-business-news-stock-market-and-financial-advice>.

15 D Hummels, 'Transportation costs and international trade in the second era of globalization', *Journal of Economic Perspectives*, vol. 21(3), 2007, pp. 131-154.

16 'Rothschild family', *Wikipedia, The Free Encyclopedia*, 2015, retrieved 22 February 2015, <http://en.wikipedia.org/wiki/Rothschild_family#The_Napoleonic_Wars>.

17 N Gershenfield, 'Digital reality', Edge.org, retrieved 22 February 2015, <http://edge.org/conversation/neil_gershenfeld-digital-reality>.

18 F Bacon, *The Advancement of Learning*, 6th edn, Macmillan, London, 1895.

19 R Harper, et al (Eds), *Being Human: Human-Computer Interaction in the Year 2020*, Microsoft Research Ltd., Cambridge, 2008.

20 J Thompson, 'Is nonverbal communication a numbers game?', blog, *Psychology Today*, 2011, retrieved 18 February 2015, <https://www.psychologytoday.com/blog/beyond-words/201109/is-nonverbal-communication-numbers-game>.

21 'When ideas have sex', video recording, Presenter M Ridley, TEDGlobal, 2010, retrieved 18 February 2015, <http://www.ted.com/talks/matt_ridley_when_ideas_have_sex>.

22 M Ridley, 'When ideas have sex', *Design Mind*, Frog Design, 4 November 2010, retrieved 15 March 2015, < http://www.fastcodesign.com/1662632/frog-design-when-ideas-have-sex>.

23 D Tuffley, 'In 10 years, your job might not exist. Here's how to make sure you're still employable', *The Washington Post*, 5 January 2015, retrieved 18 February 2015, <http://www.washingtonpost.com/posteverything/wp/2015/01/

05/in-10-years-the-job-market-will-look-totally-different-heres-how-to-make-sure-youre-ready/>.

[24] *Oxford Dictionaries*, s. v. 'robot', Oxford University Press, retrieved 7 February 2015, <http://www.oxforddictionaries.com/definition/english/robot>.

[25] M J J Lommen, I M Engelhard & M A van den Hout, 'Susceptibility to long-term misinformation effect outside of the laboratory', *European Journal of Psychotraumatology*, 2013, retrieved 18 February 2015, <http://dx.doi.org/10.3402/ejpt.v4i0.19864>.

[26] D Robson, 'Memory: Remembrance of things to come', *New Scientist*, issue 2885, October 2012, pp. 32-33.

[27] BC Storm and S M Stone, 'Saving-enhanced memory: The benefits of saving on the learning and remembering of new information', *Psychological Science*, 9 December 2014, retrieved 18 February 2015, <http://pss.sagepub.com/content/early/2014/12/09/0956797614559285.abstract>.

[28] 'Enterprise Findability without the Complexity', Google, 2008, retrieved 15 March 2015, <https://static.googleusercontent.com/media/www.google.com/en/US/work/enterprise_solutions/pdf/findability-wp.pdf>.

[29] S Ismail, *Exponential Organizations: Why New Organizations are Ten Times Better, Faster and Cheaper than Yours (And What to Do about It),* New York, Diversion Books, 2014.

[30] 'Let's use video to reinvent education', video recording, Presenter S Khan, TED, 2011, retrieved 18 February 2015, <http://www.ted.com/talks/salman_khan_let_s_use_video_to_reinvent_education/transcript>.

[31] For more on this check out A Gawande, 'Personal Best: Top athletes and singers have coaches. Should you?,' *The New Yorker*, 3 October 2011, retrieved 18 February 2015, <http://www.newyorker.com/magazine/2011/10/03/personal-best>.

[32] S Godin, S, 'A shortage of digital coaches', blog, TypePad, 2008, retrieved 11 February 2015, <http://sethgodin.typepad.com/seths_blog/2008/01/a-shortage-of-d.html>.

[33] *Dictionary.com Unabridged*, s.v. 'habit', Random House Inc., retrieved 11 February 2015, <http://dictionary.reference.com/browse/habit>.

[34] K Kelly, 'The technium', Edge.org, retrieved 11 February 2015, <http://edge.org/conversation/the-technium >.

[35] Harper, et al.

[36] C Thompson, 'The joy of typing: How racing along at 60 words a minute can unlock your mind', *Medium*, retrieved 11 February 2015, <https://medium.com/message/the-joy-of-typing-fd8d091ab8ef>.

[37] C A Christensen, 'Relationship between orthographic-motor integration and computer use for the production of creative and well-structured written text', *The British Journal of Educational Psychology*, Dec 2004, vol. 74(Pt 4), pp. 551-64.

[38] *Dictionary.com Unabridged*, s. v. 'cyborg', Random House Inc., retrieved 11 February 2015, <http://dictionary.reference.com/browse/cyborg>.

[39] L Stone, 'Continuous partial attention', blog, Linda Stone, retrieved 22 February 2015, <http://lindastone.net/qa/continuous-partial-attention/>.

[40] I Bogost, 'Cow clicker: The making of obsession', blog, Ian Bogost, retrieved 20 February 2015, <http://bogost.com/blog/cow_clicker_1/>.

[41] King Digital Entertainment plc, 'King reports fourth quarter and full year 2014 results, announces special dividend of $0.94 per share', press release, King.com Ltd., February 2015, accessed 20 February 2015, <http://investor.king.com/investors/news/financial-releases/Press-Release-Details/2015/King-Reports-Fourth-Quarter-and-Full-Year-2014-Results-Announces-Special-Dividend-of-094-per-Share/default.aspx>.

[42] Nevada Gaming Control Board, 'The average big Las Vegas strip casino, 2014', press release, University of Nevada, accessed 20 February 2015, <http://gaming.unlv.edu/reports/bigstripcasino2014.pdf >.

[43] Statista, 'Time spent playing games by social gamers in the US in 2011', Statista Inc., retrieved 20 February 2015, <http://www.statista.com/statistics/202839/time-spent-playing-games-by-social-gamers-in-the-us/>.

[44] M H Goldhaber, 'The attention economy and the net', *First Monday*, vol. 2, 1997, pp. 4-7.

[45] F Ruscica, 'Goldhaber and attention economy', blog, TypePad, June 2005, retrieved 22 February 2015, <http://edgeperspectives.typepad.com/edge_perspectives/2005/06/goldhaber_and_a.html>.

[46] From *The Amazing Spider-Man*, #216, sourced from D Beres, 'Awesome Life Advice from your Favorite Superheroes', Reader's Digest Association, Inc., retrieved 18 February 2015, <http://www.rd.com/slideshows/awesome-life-advice-from-your-favorite-superheroes/#slideshow=slide13>.

[47] American Society of Anaesthesiologists, 'When millennial med students' iPad use for instruction goes up, personal use goes down', press release, 12 October 2013, retrieved 18 February 2015, <https://www.asahq.org/newsroom/news-releases/2013/10/med-students-ipad-use>.

[48] *Oxford Dictionaries*, s. v. 'judgement', Oxford University Press, retrieved 21 February 2015, <http://www.oxforddictionaries.com/definition/english/judgement>.

[49] H Polites, 'IT policies a "secret weapon" in tech talent war: Google', Business Spectator Pty. Ltd., 7 June 2013, retrieved 20 February 2015, <http://www.businessspectator.com.au/news/2013/6/7/technology/it-policies-secret-weapon-tech-talent-war-google>.

[50] S Sreenivasan, et al, 'Social consensus through the influence of committed minorities', *Physical Review E*, vol. 84(1), 2011.

YOU'RE STILL HERE?

Well this is a little bit awkward. Clearly you want to keep reading and the book has kind of finished. Don't worry though, there is a lot more great content on my website. You can find it by visiting simonwaller.com.au. You can even sign up to my Digital Insights newsletter and get new things to read delivered directly into your inbox.

Please also feel free to contact me on simon@simonwaller.com.au to tell me what you thought of the book and how it has helped you. I also speak and do some consulting and mentoring around this stuff so if you think I might be able to help you with a project you're working on please reach out.

This really is the end now. It's time to go.

Made in the USA
Monee, IL
13 September 2026

62979533R00121